SOILLESS CULTURE - USE OF SUBSTRATES FOR THE PRODUCTION OF QUALITY HORTICULTURAL CROPS

Edited by **Md Asaduzzaman**

Soilless Culture - Use of Substrates for the Production of Quality Horticultural Crops
http://dx.doi.org/10.5772/58679
Edited by Md Asaduzzaman

Contributors
Jamal-Ali Olfati, Giancarlo Fascella, Fusheng Ma, Haiyan Fan, Jule Jankauskiene, Usama El Behairy, Md Asaduzzaman, Md. Saifullah, Md. Mokter Hossain, Akm Salim Reza Mollik, Golam Morshed Abdul Halim, Toshiki Asao

CBS, Edition **2017**

Published by InTech
Janeza Trdine 9, 51000 Rijeka, Croatia

Publishing Process Manager Danijela Duric
Technical Editor InTech DTP team
Cover InTech Design team

Additional hard copies can be obtained from orders@intechopen.com

Soilless Culture - Use of Substrates for the Production of Quality Horticultural Crops,
Edited by Md Asaduzzaman
p. cm.
ISBN 978-953-51-1739-1

INTECH

open science | open minds

Contents

Preface

Soilless culture is the modern cultivation system of plants that use either inert organic or inorganic substrate through nutrient solution nourishment. It is possibly the most intensive culture system utilizing all the resources efficiently for maximizing yield of crops and the most intense form of agricultural enterprises for commercial production of greenhouse vegetables. In fact, this an alternative cultivation system to traditional have been introduced to avoid soil related problem such as soil exhaustion, soil-borne diseases, secondary salinity development and to improve plant growth condition such as temperature and aeration of root zone, optimal distribution of water and nutrients, and also to reduce the amount of labor needed. This protected cultivation system can control the growing environment through management of weather factors, amount and composition of nutrient solution and also the growing medium. Therefore, quality of horticultural crops grown through soilless culture improves significantly compared to conventional soil culture. Production or biosynthesis of bioactive compounds will largely be depending upon the manipulation of physical, chemical and biological characteristics of the substrates that must correlate with water and fertilizer supply, climatic conditions and plant demand. Various modification of pure solution culture has been taken place over time throughout the world. Primarily, gravel or sand was used in soilless culture system to provide plant support and retain mineral nutrient and water. Afterward, several substrates have been used for their unique properties of holding moisture, aeration, leaching or capillary action, and reuse potentiality. Organic substrates includes sawdust, coco peat, peat moss, woodchips, fleece, marc, bark etc. whereas, inorganic substrate of natural origin are perlite, vermiculite, zeolite, gravel, rockwool, sand, glass wool, pumice, sepiolite, expanded clay, volcanic tuff and synthetically produced substrates are hydrogel, foam mates or polyurethane, oasis or plastic foam etc.

This book is targeted at commercial vegetables and ornamentals growers who are thinking of increasing the efficiency and quality of their agricultural produce through modifications of the environmental controls, management of culture systems and use of technological innovations. It describes the design and preparation of nutrient solution used for soilless culture, influence of growing substrate on physiological processes, yield and quality of vegetables. This work is also devoted to describing growing substrates alternative to peat for ornamental plants, suggesting substrate water scarce area with key technologies of increasing irrigation efficiency of successful crop production. This book will provide much valuable information for the commercial growers, researchers, and the students of the field.

The publication of this book would have been impossible without the dedication and hard work of many researchers around the globe. All acknowledgements go to the authors of these chapters, who volunteered their valuable time to contribute to this book.

Dr. Md. Asaduzzaman
Olericulture Division, Horticulture Research Centre,
Bangladesh Agriculture Research Institute,
Joydebpur, Gazipur-1701, Bangladesh

Chapter 1

Influence of Soilless Culture Substrate on Improvement of Yield and Produce Quality of Horticultural Crops

Md. Asaduzzaman, Md. Saifullah,
AKM Salim Reza Mollick, Md. Mokter Hossain,
GMA Halim and Toshiki Asao

Additional information is available at the end of the chapter

http://dx.doi.org/10.5772/59708

1. Introduction

Soilless culture is the modern cultivation system of plants that use either inert organic or inorganic substrate through nutrient solution nourishment. Possibly it is the most intensive culture system utilizing all the resources efficiently for maximizing yield of crops and the most intense form of agricultural enterprises for commercial production of greenhouse vegetables [1-3]. Several studies suggested soilless culture in the greenhouse as an alternative to traditional field production for high-value vegetable crops [4-7]. This protected cultivation system can control the growing environment through management of weather factors, amount and composition of nutrient solution and also the growing medium. Therefore, quality of horticultural crops grown through soilless culture improves significantly compared to conventional soil culture [8,9]. This artificial growing system provides plants with mechanical support, water and mineral nutrient for higher growth and development. Over the years, hydroponics has been used sporadically throughout the world as a commercial means of growing both food and ornamental plants. Now at days, it has also been used as the standard methodology for plant biological researches in different disciplines [10]. Various modification of pure solution culture has been taken place over time throughout the world. Primarily, gravel or sand was used in soilless culture system to provide plant support and retain mineral nutrient and water. Afterward, several substrates have been evolved due to their unique properties for holding moisture, aeration, leaching or capillary action, and reuse potentiality. Soilless growing media are easier to handle and it may provide better growing environment (in terms of one or more aspects of plant growth) compared to soil culture [11,12]. Organic substrates includes sawdust, coco peat, peat moss, woodchips, fleece, marc, bark etc. whereas, inorganic substrate of natural origin are perlite, vermiculite, zeolite, gravel, rockwool, sand, glass wool, pumice, sepiolite, expanded clay, volcanic tuff and synthetically produced substrates are hydrogel, foam mates (polyurethane), oasis (plastic foam) etc. [13-18]. Various raw materials have been used to produce growing media for vegetable production throughout the world. Capabilities of compost for use in soilless culture of horticultural crops have also been confirmed in a number of studies [19,20]. Oil palm frond compost has a great potential to be utilized for the improvement of soilless culture system. It is successfully used to control plant diseases [21-24]. In addition, plant nutrients from oil palm frond compost are released slowly over a long period of time and are less likely to leach out of the media.

Although successful cultivation of different vegetables and ornamentals crops in soulless culture with bark source have been reported [25-27], phytotoxicity from phenolic compounds may be extracted from the substrate [28,29]. Therefore, it is evident that at present, utilization, standardization of nature of raw materials used for soilless growing media is diverse in origin [30]. Each substrate has its specific properties and usually differs from others. These differences between growing media have to be considered for successful soilless cultivation of horticultural crops. In this instance Gruda *et al.* [31] suggested the activity of microorganisms must be evaluated in comparing peat and its substitutes, such as bark, wood fiber substrate, paper and straw substrates. In order to build up own body protein components, these microorganisms need mineral nitrogen, which they gain from the available nitrogen content in the substrate. Therefore, nitrogen would not be readily available for the plants in soilless substrate which in turns may lead to potential quality losses of the produce [31].

In recent years, the use of soilless culture has increased significantly throughout the world [2, 32]. More than 60% of the vegetable greenhouses in the Netherlands cultivated using rockwool media but it is costly and difficult to dispose because it is not biodegradable and environmental friendly [26,33]. Perlite which is less expensive than rockwool has been used as soilless culture substrate around the world for successful production of vegetables, fruits, and cut flowers in the greenhouse [2,34]. Similarly, zeolite has also the potentiality as soilless media for its unique properties. Zeolite crystal alumina silicates have negative charges, which is balanced by one or two valence of positively charged cations [35]. It has high water absorption, retention and releasing capability, high cation exchange capacity, and high buffering ability of pH change [36]. It has been found that due to its higher cation exchange capacity, water and nutrient holding ability; yield and fruit quality of tomato increased greatly [37]. Soilless culture of gerbera produced higher yield in perlite/zeolite (1:1) substrate than other mixtures, due to its improved aeration and water retention ability [38]. Another substrate is coconut coir has a great demand by the ornamental industries especially in The Netherlands and Canada [39], and more recently, the product has been marketed as a substitute for rockwool in the greenhouse vegetable industry. There are many indigenous and locally available soilless culture substrates used by different countries and similarly produced synthetic substrate suitable for growing system of specific crops. Use of different locally available and inexpensive soilless substrates with no pollution limitations but with adequate physical and chemical properties has been suggested worldwide. Mixture of different substrates also been used for higher growth and yield of several crops around the world [40-43]. Soilless culture in bags, pots or trays with light weight medium is the simplest, easiest and economical way of growing crops. The most common types of growing media in container based system are peat-lite, a mixture of bark and wood chips [44]

The problems in agricultural land use such as soil exhaustion, pest infestation or chemical interference are increasing greatly due to intensive cropping, injudicious application of pesticides or continuous monoculture [45-49]. In this regard, soilless culture can avoid problems with monoculture of plants in the same land for years [50]. It can provide several major advantages in the management of both plant nutrition and plant protection. The main reason of need for soil to soilless culture for horticultural crops is the problem related to proliferation of soil borne pathogen in the soil cultivation. Research studies reported that commercial production of greenhouse vegetables with soilless media adopted to reduce economic losses caused by soil-borne pathogens [51-53]. While other researchers reported that soilless culture can provide more efficient use of water and fertilizers [54,55], reduce root diseases [56], and facilitate cultivation of crops in areas where normal cultivation is not possible [57]. Thus, soil has been replacing by many organic and inorganic substrates, since they are disease and pest free inert material capable of holding required sufficient moisture and can be reused year after year. The physical and hydraulic properties of soilless culture substrate is better that those of soil medium. In soil culture plant root get higher water availability just after irrigation which cause lower oxygen content to be used by plant root and micro flora but in substrates optimum aeration is possible due to its leaching or pulling capacity by capillary action. Water application is several times higher in tomato (4 times) and lettuce (5 times) under conventional cultivation system compared to hydroponics [58]. Root development and nutrient absorption is less in plants grown in soil but soilless substrates especially inorganic origin can hold adequate moisture, nutrient through their surface charge and also allow profuse root hair formation for efficient absorption. However, root volume is restricted in container based substrate culture. This limitation has several beneficial effects such as limited supply of nutrient is possible in soilless substrate culture [59, 60] and also increases the root to root competition since there are more roots per unit volume of medium.

Substrate culture under protective agriculture has minimized the discharge of fertilizer and pesticide residues into the natural environment such as freshwater reservoir. However, there are several observations to be considered for successful crop grown in soilless substrate culture. The limited volume of substrate and water availability can cause rapid decrease in water and mineral nutrient status. Therefore, changes in amount of solution, its electrical conductivity (EC), and pH should be monitored regularly for efficient use of water and nutrients. In soilless substrate mineral nutrient usually supplied as ionic form and thus when plant exposed to low relative humidity, it lose water by transpiration leading to evaporation of water from the medium and plant tissue. This transpiration and evaporation can lead to salt build-up in the substrate due to improper management.

Suitability of different substrates in successful vegetable establishment and their effect on growth, yield and produce quality have been extensively investigated by many researchers around the world. However, only few researches have been conducted for improvement of horticultural crop quality in different substrates. Recent reviews suggested that changes in quality parameters of horticultural crops influenced by the use of growing substrate [152] and present a comprehensive overview of the effect mineral soil, inorganic and organic growing media on the growth, development, yield and quality of vegetable crops grown under greenhouse condition [13]. This chapter aims mainly to describe the importance of soilless culture for enhancing quality production of horticultural crops, improving produce quality beneficial to human health, economics of reutilization of once used substrates and also the prospect of soilless culture in improving and maximizing crop yield.

2. Improvement of horticultural produce quality through soilless hydroponics

Horticultural produce from soilless culture have better qualities than those from conventional soil-based cultivation [8,61-63]. Although the exact differences between qualities of vegetables grown in soil or hydroponics are difficult to determine [64] but soilless culture in greenhouse may be an alternative to soil culture for high-value vegetables crops including tomatoes, peppers, cucumbers, lettuce etc. In a study, Massantini *et al.* [9] found better taste, uniformity, color, texture and higher nutritional value in fruits grown in soilless culture than in soil cultivation methods. Similarly, it was also found that tomatoes produced in the nutrient film technique system were firmer and richer in vitamin C than those grown from soil-based plants. It also contained more sugar, acid and sodium, resulting in a distinct taste. Vegetables from organic substrate culture in greenhouse and poly tunnels are in high demand. Thus, in order to increase the qualities of horticultural produce appropriate fertilizer application, especially nitrogen and phosphorus along with growing substrate prepared from organic materials are suggested [65]. Several studies showed that in general plants harvested from soilless culture had a lower dry weight and leaf area, however, significantly higher productivity were observed at the end of harvest [66]. In this culture system, high concentration of nitrogenous fertilizer enhance the vigorous growth, which reduce the penetration of light intensity to the whole canopy due to huge foliage and thus reduce the accumulation of ascorbic acid in shaded parts. Enhanced growth of plants due to nitrogenous fertilizer may also have a relative dilution effect in plant tissue. Therefore, excess use of nitrogenous fertilizer increases the concentration of nitrate in plant tissue and simultaneously decreases that of ascorbic acid, it may have double negative effect on the quality of plant foods [67].

In a study it was found that, potassium concentration in plant parts may vary for growing seasons (spring or autumn) and also growing systems [68]. It was reported that tomato plant grown in aeroponics gave higher concentrations of P, K and Mg and lower concentrations of Ca than nutrient film techniques [69]. Substrate culture found to be affected greatly increasing mineral contents in plants especially due to luxurious nutrient uptake during vegetative growth [70]. Fruit quality of tomato is greatly influenced by potassium mineral nutrition. It positively affects the contents of soluble sugars, vitamin E, carotenoids in fruits but its luxurious absorption may also negatively affect the uptake of magnesium, calcium, and boron from nutrient solution [71-73]. This antagonistic interaction of potassium with calcium leads to decrease in concentration of calcium in the medium. As a result, a typical symptom generally appears known as blossom-end rot disease on tomato fruit which lower the quality greatly [74]. Despite application on the same medium, various substrates like sand, mine material of volcanic origin, rockwool, wood fiber, peat and coir showed significant differentiation in the nutrient content [75-78].

Soilless culture has been extensively used in tomato cultivation both in commercial and experimental basis. Many researchers has compared, standardized and otherwise applied various substrates in tomato culture in soilless hydroponics. In general soilless culture reported to increase the tomato fruits quality greatly around the world. It has been found that organic growing media produced higher yield and number of fruit than conventional growing system in greenhouse tomato production [13,79]. Many studies also suggested that tomato fruits grown in organic substrates had higher dry matter, vitamin C, and nitrogen compared to rockwool [80,81]. Similarly, these properties were improved in rape straw substrate along with peat and pine bark compared to rockwool [82]. The quality and quantity of tomato fruit in organic media found better than inorganic media [83] and when it grown in different substrates the highest amount of total yield and number of fruits were harvested from perlite + rice hull while fruits with highest total soluble solids were from coco-peat substrate [84]. Tomato plants grown in perlite and zeolite mixture substrate (2:1) produced greater fruit size, total soluble solid, sensorial qualities and also highest dry matter of fruit [85] and it was also reported that cucumber plants grown in nutrient film technique gave higher fruit quality than plants grown in perlite culture [86]. Fruit qualities such as fruit weight, fruit firmness, total soluble solids, titratable acidity, ascorbic acid and carotenoids were found to be influenced by the soilless substrate used, while they had not any effect on EC, pH and dry matter content.

Utilization of rockwool and perlite in soilless hydroponic culture results in higher yield compared to other inert materials [84]. However, it also reported that tomato grown in substrate prepared from cutting pieces of rye and wheat straw [17] or slabs made of shredded rye straw [88] yielded higher than that from rockwool cultivation. The tomato plants that grown in perlite and zeolite with 2:1 ratio had best distribution of fruit size, total soluble solid and sensorial quality and highest dry matter of fruit was found in perlite substrate [85]. Research results also suggested that addition of maize to perlite and pumice could improve properties of inorganic substrates for tomato soilless culture, leading to higher yields and better quality fruit [87]. Most of the sensory characteristics such as redness of surface skin, firmness, crispness, sourness, sweetness, tomato aroma and overall impression after chewing were varied greatly due to differences in variety, followed by maturity, harvest time and EC but type of growing medium either soil or rockwool had no or little effect. However, for the characteristics related to texture (crispness and firmness), the ranking was harvest time, EC, growth medium, maturity and variety, with soil-grown tomatoes being slightly but significantly softer than the rockwool grown tomatoes [89]. Higher EC values in the growing medium may cause decrease in fruit yield but on the other hand, it improves the taste by increasing dry matter, soluble solids, and titratable acidity [90]. It has been found that salinity of the water improves the quality of tomato [91]. In soilless culture, increase the EC value of irrigation water or that of nutrient solution increase the acidity [92], the soluble sugars [93,94], and dry matter percentages of fruits [92,95] while decrease the size of fruits in cherry tomato [96].

Maize stems having light weight and less costly can be used as substrate in soilless culture which contains readily available organic matters [97]. In another study, tomato fruit quality characteristics such as mean fruit weight, fruit firmness, total soluble sugars, titratable acid, carotenoids, and ascorbic acid were affected differently by the use of maize shredded stems, perlite and pumice substrates and among them maize shredded stem substrate resulted in greater fruit firmness compared to perlite, pumice substrate [98]. Customer tests indicated that firmness and flavor are important criteria for high quality tomato, where typical tomato flavor depends on the ratio between sugar and acid [99]. Higher sugar and organic acid content improves the quality of tomato fruits [100]. Amount of citric acidity in tomato fruits was found as higher or similar in tuff or sand substrate compared to soil medium [101]. In another study with lettuce, Siomos *et al.* [102] found that soilless culture results in higher citric acid percentage compared to soil culture. However, fruit size and quality characteristics also showed no significant difference within substrate of coco-peat, rockwool and masato [103]. Harvesting time of tomatoes had influence on the quality parameters as in September harvest produced higher dry matter and carotenoids content than that of June harvested fruits. However, June harvested fruits were characterized by a higher total sugars content, pH of juice and soluble solids content [104]. In this regards, it is mentionable that tomatoes sensory quality mainly determined by sugar content which represent the major components of soluble solids [105].

After tomato lots of research works have been conducted on soilless substrate for its influence on improvement of growth, yield and quality of pepper. Growing media composed of soil, peat, perlite, sand and pumice significantly affect the yield, fruit weight, ascorbic acid values and total soluble solids of pepper cultivars [106]. The highest early yield was obtained in pepper plants grown on the peat medium compared to perlite, pumice, sand and soil [106]. Schnitzler *et al.* [107] observed better plant growth, fruit yield and quality in bell pepper (*Capsicum annuum* L.) grown in wood fiber substrate. Recent studies showed that plants grown on peat media had higher ascorbic acid content, total soluble solids, fruit number per plant and yield than its mixture with perlite or sand [108]. Peat contains higher potassium than its mixture substrates [108] and it has been reported that growing media with high potassium could increase the vitamin C content in plants [109]. Green peppers were grown in mixture of substrates such as vermiculite + sand, peat + perlite and rockwool showed that peat + perlite had most influence on its growing traits and yield [110]. However, when perlite compared with rice husk substrates it was found that plants grown in rice husk had higher growth and yield in the later [111]. In another study, differential response of growing substrates were reported and they showed significant effect on plant height, number of leaves, chlorophyll index and total yield per plant [112].

In strawberry better growth has been reported in coir than that in perlite substrate [113]. In another study, the influence of different substrates on the growth of strawberry was reported as peat, finpeat or finpeat + perlite in Camaros and Fern cultivars [114]. Jafarnia *et al.* [115] reported total soluble solid were influenced by substrate and cultivars and fruit qualities such as vitamin C and titratable acidity were highest in rice husk substrate. Caso *et al.* [116] used rice husks and pumice with different ratios in column system for the production of strawberry and they recommended that 100% rice husks substrate influence majority of measured traits.

It also found that content of phenolic compounds, especially anthocyanin depend on substrate pH [117] while Lopes da Silva *et al.* [118] reported total anthocyanin would range between 200 and 600 mg kg^{-1} fresh weight. From research results it is evident that soilless culture substrate affect the quality of strawberry and desirable fruit production is greatly depends on suitable choice of substrate and cultivars [119]. They found that highest total anthocyanin content and titratable acidity in Camarosa cultivar in vermiculite + perlite + coco-peat; the highest antioxidant in Camarosa and Mrak cultivars in substrate of Sycamore pruning waste and coco-peat + perlite; and the highest total soluble solids in Selva cultivar in vermiculite + perlite + coco-peat substrate. Strawberries grown in greenhouses with different soilless growing media also showed their impact on phytochemical and nutritional composition [120]. Agricultural cropping systems greatly influence the productivity and yield of crops. It has been reported similar [121,122] or even higher [123] yield for organic corps than conventional soil cultivation. Minerals such as calcium and magnesium concentrations were observed higher in organic and low input soil system but soilless growing system produced fruits with higher firmness in the green stage which is related to higher flesh thickness of fruits [124].

Rockwool substrate can be used to produce melons hydroponically [125,126] but costs would be higher than other substrate materials and its disposal is very difficult [127,128]. Recently, Rodriguez *et al.* [129] investigated different combinations of media (coarse and medium perlite) and containers (polyethylene bags and plastic pots) for hydroponic production of 'Galia' muskmelons (*Cucumis melo* L.) and found that fruit yield and quality were not affected by any combination of media and containers. In recent studies it was found that sweeter cantaloupes or rock melon fruits harvested in plants grown in empty fruit branch media than coconut dust as soilless media [130]. Effect of different substrates has been studied on growth, yield and quality of watermelon in soilless culture [131]. Quality and quantity of watermelon fruit had not any significant difference between different substrates evaluated [131]. Influence of peat substrate and its mixture with perlite or zeolite on the quality of cucumber seedlings and photosynthesis parameter has investigated [132]. It has been suggested that the highest yield of cucumber fruit obtained from cocopeat substrate than other substrates like perlite-cocopeat (50-50, v/v), perlite-cocopeat-peatmoss (50-20-30 and 50-30-20, v/v) and other growth indices such as stem diameter, biomass, fruit's number, fruit size and fruit diameter were greater in cocopeat [133]. In another study, it was showed that total soluble solids along with growth indices such as yield, biomass weight, shoot diameter, plant height, root weight, and leaf area index of cucumber plant were significantly higher in date-palm substrate than soil media but generally had no significantly difference as compared with perlite substrate [134].

In a recent study, carrots were grown successfully in hydroponics using perlite substrate [135]. It was found that carrot plants grown in 0.6 mm perlite supplied with 100% nutrient solution produced significantly higher root yield compared to larger perlite particles and higher concentrations of nutrient solution. Carrots grown in 0.3 mm perlites produced shorter roots, wider near the proximal end and whitish in the distal end due to excessive water content causing oxygen deficiency. It was found that seedlings grown in peat substrate are higher, have bigger leaf area than seedlings grown in peat-perlite, peat-zeolite substrate, but in leaves and roots dry matter accumulation was less. Higher tuber yield in potato grown in hydroponics compared to conventional system was reported [136]. This higher tuber yield was attributed by the uninterrupted and optimal nutrient and water supply in hydroponic culture.

Soilless culture has predominant influence on the floriculture industries and can provide means of best quality flowers production throughout the year. In roses industry, higher yield and best quality of stems are entirely depends on physico-chemical properties of growing substrates. It was found that incorporation of rice hulls and press mud in traditional substrate found to be improved the growth and quality indices and increased flower yield of *Rosa hybrids* L. cvs. 'Kardinal', 'Anjlique' and 'Gold Medal' [137]. Fascella and Zizzo [138] studied that soilless cultivation of roses in perlite or coconut coir dust increased yield and stem quality. This might be related to the higher water holding capacity and cation exchange capacity of coconut coir, suggesting this organic substrate is one of the alternatives to peat for hydroponic culture. The highest quality of cut flowers of gypsophila in terms of stem length and number of branches per flower were obtained from plants grown in sawdust growing medium under soilless hydroponics with bag culture [139]. High quality cut flowers of oriental hybrid lily were obtained in solid medium hydroponics when compared to mist culture system [140]. It was also observed that broken chaff substrate induced higher quality lily cut flowers as compared with chaff, hydro-ball or carbonized chaff substrate. Hsu *et al.* [141] grew *Oncidium* orchids in rockwool, sphagnum peat moss and mixed medium containing crushed stone, bark and charcoal. They found that pseudo bulbs mass, root activity, cut flower qualities in terms of flower length, floret number and number of shoots were higher in rockwool compared to other media. However, little difference in yield and quality could be attributed due to types of soilless medium used under adequate management practice and environmental conditions [126,142]. The amount of nutrients in both organic and inorganic substrates changes during active vegetative growth of plants and its indication may be appeared in the leaves. Thus frequent analysis of substrate, at least once a moth is important for successful cultivation under soilless cultivation [76,78,143].

3. Production of specialty crops providing human health benefits through soilless hydroponics

The world's population increased greatly in last few decades. The improvement of living standard in many countries increased with the great demand for high value crops, off season supply and high quality products. Therefore, quality of life (QOL) of people increased considerably. In this regard, protected agriculture which is a labor intensive industry can produce higher amount of food for the increased population of the world. The efficiency and quality of the agricultural produce can be increased through the modifications of the environmental controls, management of culture systems and use of technological innovations. The greatest advantage of soilless culture is that it allows direct control of the nutrient solution, possible to modify composition and concentration to achieve predictable results in relation to dry matter content, nitrate content or other organoleptic and structural features of the crop produce [144]. Thus, physical, chemical and biological characteristics of the substrates must correlate with water and fertilizer supply, climatic conditions and plant demand [145-149]. In addition, production or biosynthesis of bioactive compounds will largely be depending upon the manipulation of these characteristics. Phenolic acids are important bioactive compounds having antioxidant activity. Tomato fruits are the good source of phenolics usually taken by human through their daily diet [150,151]. However, it was found that growing medium (standard mineral wool slabs or coconut fibre slabs) or harvest term (September or June) had no influence on the phenolic acids content in the tomato fruits [104]. Other studies also showed that the qualitative traits of the products obtained from soilless culture appear to be substantially similar to the products coming from conventional cultivation [152,153]. Soilless culture may improve the parameters related to nutritional, organoleptic and hygienic-sanitary characteristics [152,153] but some aspects of vegetable quality reported to be clearly improved, such as phytosanitary residues, enhanced organoleptic characteristics and longer shelf life [154]. Special dietary requirements are also sometime fulfilled e.g., enrichment of and/or increase in selenium [155], iron [156], omega 3 [157], and lowering the nitrate [158], and potassium content [159].

Soilless substrate originated from organic materials would improve the product quality with health promoting substance. Many studies indicated that higher nutritional value and higher content of biologically active compounds in the agricultural products from organic farming [160-162]. However, other studies reported that effect of cultivation method disappears when the results converted to absolutely dry matter [163,164]. In most studies it also found that vitamin C content in organic fruits is higher than that of conventional tomatoes [163-165]. In conventional cultivation methods, tomato plants absorb easily assailable nitrogen from the substrate. A large concentration of this macro element results in increased synthesis of protein components and proteins, which adversely affect the synthesis of carbon-based compounds such as vitamin C. Therefore, plant products from organic farming are higher in vitamin C compared to conventional system [166,167]. Organic growing system also influence the nutritional value and phenolic compound content in tomato [168] and a two years study showed that organic tomato had higher ratio of reducing sugar/organic acids, more total sugars, vitamin C, total flavonoids, 3-quercetin rutinoside, quercetin-3-*O*-glucoside, myricetin, chlorogenic acid and kaempferol content than convention fruits.

Research reports revealed that tomato flavor is related to the balance between total soluble sugars and organic acids in the fruits [169]. It has been found that potassium fertilization had positive effect on fruits sugar and acid content [170], therefore, soilless substrate containing higher amount of potassium will increase the sweet flavor of fruits. Potassium supplied from the growing media also influences the antioxidant content of tomatoes, which is considered as beneficial for human health. On the other hand vitamin C is a health-promoting substance with antioxidant properties, which in turn play efficient role in preventing the conversion of nitrate to nitrite in plant tissue and within the human body [171]. Amount of nitrogen absorbed is an important factor influencing the vegetables quality and the way in which absorbed nitrogen is utilized in plant metabolism either as nitrate or nitrite form in the edible plant tissue [172]. These factors can be better managed in hydroponics through management and supply of nutrient solution composition in the small volume of rooting or culture medium.

Fast growing fruits and leafy vegetables had great potential for enrichment of minerals, bioactive compounds and health promoting substances. Commercial cultivation of these crops for a specific dietary requirement can be possible in order to meet the demand of such type of people. Cultivating leafy vegetables in a floating system is the easiest and cheapest means of production, since this system shows high water and nutrient use efficiency with low environmental impact [173]. This cultivation system produced acceptable yield and good control quality parameters in baby leaf species. Siomos *et al.* [174] found that plants from a soilless culture had higher nitrate, total nitrogen, phosphorus and potassium content compared to plants harvested from soil culture. Fruits and vegetables grown in soil contaminated with environmental toxicant or pollutant from industrial effluent or heavy pesticide application have higher mineral contents along with toxic heavy metals and if accumulated in their tissue will impose potential health risk to human [175-177]. Surface soil act as toxic chemical filters that may absorb and retain toxicants from waste water and other effluents. However, due to continuous accumulation of these pollutants and changes in soil pH, the capacity of soil to retain toxic elements reduces and thus surface soil permit these elements to pass into ground water or available for plant uptake [178]. Micronutrients and heavy metals are a group of non-biodegradable elements with the tendency of bioaccumulation in living systems causing serious health problem [179-182]. Moreover, research results reported that some heavy metals such as Cu, Cr, Ni, Zn, Fe etc. at low doses are essential for plants but at high doses cause metabolic disorders and growth inhibition especially Pb and Ni [183,184].

Industrial effluent often contains considerable quantities of heavy metals and other substance that may be toxic to people but beneficial for horticultural crops. Therefore, it is imperative that before effluent can be used for commercial production of vegetables and fruits, it must be determined whether there is or not accumulation of heavy metals [185]. In a study, application of recycling water in broccoli caused an increased yield but it also resulted in enhanced heavy metals in tissues [186], therefore, when applying recycle water, the amount of heavy metal must be considered and managed to a minimal level. In this regard, Emongor *et al.* [187] reported that applying secondary treated sewage effluent enhanced yield of tomato when compared to the plants irrigated with tap water. Recycle water is the easily available source of nutritional supplement necessary for crop growth and thus it has reported to an increase in agricultural crop productivity [186,188,189]. Although wastewater and sewage effluents had beneficial effect on horticultural crops, it contains a significant amount of trace elements and other toxicant that are harmful to human [190]. Previously, an enhanced amount of minerals with applying recycle water has been reported [191]. Similarly, it has been reported that with application of recycle water in cabbage the amount of mineral caused an increase in tissue and resulted in enhanced yield [192]. Moreover, from the economic viewpoint, recycle water in irrigation of crops under proper agronomic and water management practices may provide higher yields and save additional cost of water and fertilizer [189].

Now a day's expensive pesticides application in controlling pests and diseases is a prerequisite for successful production of horticultural crops. Pesticide residues in the agricultural products often cause health hazards. Therefore, growing demand of high quality of fruits and vegetables with minimal or without pesticide residue is desirable to the local consumer and also for commercialization. In this case, soilless culture is a good alternative method of quality crop production [193]. Therefore, soilless culture techniques could be applied to grow selected and popular local horticultural crops with the application of food safety standards at a reasonable price [194]. In addition, injudicious use of nitrogenous fertilizers lead to the production of green vegetables with higher NO_3^- content, which considered to be cancerous to human health. Apart from soil culture, solution culture also produces vegetables with higher NO_3^- and this hazardous ion could be reduced to a greater extent through eco-organic soilless culture system [195]. In regards to NO_3^- content of fruit, the highest value was found in organically grown green peppers and the lowest values were observed in red peppers regardless of organic, low-input and soilless systems [124].

4. Reuse of soilless culture substrate with an economic view point and environmental issues

Substrate culture is considered to be main soilless technique for commercial scale production of horticultural crops. However, it has disadvantage of disposal of growing substrate after crop cultivation. In general, hydroponics is claimed to involve a high initial capital investment and need of technical knowledge for complicated cultivation procedures. However, these problems could be resolved by using locally available materials in simplified methods and equipments. For example, farmers in Japan built their own hydroponic production system using local material which much cheaper than purchasing [196]. Reviews of several research works on the use of substrate in soilless culture showed differential influences on growth, yield and quality of crops. In addition to cultivars and horticultural management practices, growing media had great influence on the yield and fruit quality of greenhouse grown tomato [197]. It was found that plants grown in perlite produced higher total marketable yield that plants grown in either rockwool or pine bark. However, the initial costs to grow greenhouse tomato in perlite were higher than rockwool and were the lowest in pine bark. Replacing perlite substrate at every growing season of tomato was found costly [197]. Continued culture with perlite substrate without proper reconditioning, desalination and disinfection may cause medium compaction, salt built up and pest infection [198-200]. Thus increased salt concentration can reduce fruit size [201], decrease fruit number [202] or can negatively impact root and shoot growth of tomato plants [203]. Therefore, reuse techniques are necessary for sustainable soilless production with lower inputs. Recent researches suggested that perlite can be recycled and used for many years, thus reduce production cost without any negative impact on crop yield [198,199]. If rockwool substrates can be steam sterilized and reused once and then it must be disposed because of fiber break-down during steam sterilization and handling [204]. Therefore, a significant cost is associated with the disposal of rockwool substrate [205]. Disposal of used substrate create environment hazardous in the 1980s, thereafter several research efforts has been taken on modern horticultural techniques to comply with ecological mandates and bio-stability of soilless substrates [206]. As a result several new organic growing media have been suggested by many researchers around the world based on renewable raw materials.

In substrate culture, continuous recirculation of nutrient solution is difficult to maintain. Low sterilization techniques are necessary, thus rockwool slabs with drainage to waste are the most common system especially in the Netherlands [207]. In soilless culture, salt e.g., Na^+ and Cl^- accumulation in the growing media is common which may exert negative effects on salt sensitive crops [208]. Therefore, collection and sterilization technique of drainage nutrient solution are to be developed for reuse [209]. In recent years, a number of investigations have been taken on the water and nutrient balance in greenhouse-grown crops [210]. It has been clearly shown that the large excesses of water and minerals absorption lead to the emissions of N in larger extent and P to a lesser extent to the environment. Therefore, recirculation of once used nutrient solution is imperative for economic crop production hydroponically. In this regards, high EC and nutrient level in the soilless medium are necessary to meet the crop requirements at the high rates obtained under protected cultivation [211] which in turns will enhance product quality grown therein [212]. On the other hand, in soil-bound crops surface water is often used, and since it contains rather high salt concentrations, leaching is necessary to prevent salinity problems [213]. Therefore, the need of leachation, sterilization and reutilization could be the process sustainable crop production through soilless culture system. However, the cost of fertilization was found to be insignificant compared with the total production cost in greenhouse cropping [214]. Crop cultivation in reused substrates revealed both positive and negative responses compared to fresh ones. Some researchers found reduction of crop yield and/or produce quality in re-used substrate [215], while others reported no or minimal differences between new and reused substrate [216-221].

Reuse of substrate is an important option for environmental management of growing media and of soilless culture. It may increase crop profitability, although substrate costs generally constitute a small fraction of the total production costs of greenhouse and nursery crops [222]. However, breakdown of substrate materials can exert detrimental effect on crop for repeated use several years. Physical and chemical modification of both organic and inorganic substrates may also occur after one or two growing cycles, and number of growing cycles of a substrate depends on its nature and the type of crops grown. Research findings showed that generally inorganic substrates tend to last longer for example; polyurethane upto 10 years [223,224], perlite upto 3 years [225], rockwool upto 3 years [226]; and organic substrates have a shorter life upto 2-3 years due to minor biostability [227,228]. Thus, physical stability of the growing medium becomes an important issue in maintaining favorable growing conditions for the whole period [229-231]. It has been suggested that substrate volume could be reducing until 25%, without yield reduction, if irrigation scheduling is adapted to the lower water buffer. Decision on prolonged the use of substrate should be taken using new quick test for assessing the physical, chemical and phytopathological conditions before the start of new cultivation [232]. Among the soilless substrate perlite has good traits for soilless cultivation because of its high water absorption ability, high water efficiency, reuse potentiality and decrease cost of production [233].

5. Future prospects of soilless culture for maximizing yield of horticultural crops

Soilless culture technique has been used successfully in the production of difficult to grown plants. It has great opportunities to explore the inabilities of production constraints involving environmental controls. Modification of culture methods and culture environment can lead to a sustainable crop production desirable for human beings. In this regards, hydroponic production of medicinal and aromatic herbs showed a new insight towards the mass production of these plants leading to high secondary metabolites yields and qualities [234-236]. Soilless culture of medicinal herbs has many valuable advantages such as high yields, clean cultivation, year round production and production of drugs with minimum herbicide and pesticide residues [237,238]. Adequate supply of water and mineral nutrients increase the absorption and subsequently higher dry matter production both in aerial and underground parts of medicinal plants are the main advantages of substrate culture compared to field grown counterparts [237,239-241]. Therefore, successful soilless hydroponics of high value medicinal plant could be promising for pharmaceutical and food industries on meeting their high demands for *Chrysanthemum balsamita* (L.) Baill. raw materials [242].

In Southern Tunisia, the application of geothermal water in soilless culture using sand as substrate found to be much more appropriate than perlite and stone pumice substrates. In sands, plants growth was faster and gave higher marketable yield with improved fruit quality having higher acidity and sugar content [243]. Transpiration influences transport and translocation of calcium in the plant body. It has been found that, nutrient and transpiration are both important in preventing blossom-end rot disease in tomato in soilless culture. Thus, mineral nutrient level i.e., EC value should be maintain for improvement of produce quality. Studies revealed that potassium and EC have positive effects on quality of vegetable crops grown in hydroponics. On the contrary, low EC found to promote quality of cut flower. Further investigation are necessary to determine the prolong reuse of the substrates and their mixtures.

The need for soilless culture arose from plant protection issues with soil-borne pathogens and environmental regulations against groundwater pollution with industrial effluents, nitrate and pesticides. Soilless substrates either having organic or inorganic ingredients have been used as for finding suitable growing media for horticultural crop production. The types of raw material used vary according to their domestic availability in the world. Raw materials variations in different substrate influence the plant growth and development directly and/or indirectly. Thus selection of ideal substrate from various materials is imperative for productivity of each crop [244]. Lots of substrates evolved for horticultural crops production with their cultural guidelines. From them only suitable or adapted cultural guidelines will benefits the grower in successful cultivation for his produce.

Author details

Md. Asaduzzaman[1*], Md. Saifullah[1], AKM Salim Reza Mollick[1], Md. Mokter Hossain[2], GMA Halim[1] and Toshiki Asao[3]

*Address all correspondence to: asadcbt@yahoo.com; asadcbt@bari.gov.bd

1 Olericulture Division, Horticulture Research Centre, Bangladesh Agricultural Research Institute, Joydebpur, Gazipur, Bangladesh

2 Department of Horticulture, Faculty of Agriculture, Bangladesh Agricultural University, Mymensingh, Bangladesh

3 Department of Agriculture, Faculty of Life and Environmental Science, Shimane University, Kamihonjo, Matsue, Shimane, Japan

References

[1] Dorais M, Papadopoulos A, Gosselin A. Greenhouse tomato fruit quality. Horticultural Reviews 2001;26: 239-319.

[2] Grillas S, Lucas M, Bardopoulou E, Sarafopoulos S, Voulgari M. Perlite based soilless culture systems: Current commercial applications and prospects. Acta Horticulturae 2001;548: 105-114.

[3] Jensen M. Food production in greenhouses. In: Plant Production in Closed Ecosystems: The International Symposium on Plant Production in Closed Ecosystems. Kluwer, Dordrecht, The Netherlands. 1997; p.1-14.

[4] Blank C. Specialty process for specialty products. The Grower Magazine, March 1999; p.28-30.

[5] Cantliffe DJ, Shaw NL, Jovicich E, Rodriguez JC, Secker I, Karchi Z. Passive ventilated high-roof greenhouse production of vegetables in a humid, mild winter climate. Acta Horticulturae 2001;559: 195-201.

[6] Paradossi A, Malorgio F, Campiotti C, Tognoni F. A comparison between two methods to control nutrient delivery to greenhouse melons grown in recirculating nutrient solution culture. Scientia Horticulturae 2002;92: 82-95.

[7] Schroder FG. Alternative vegetables grown in hydroponic systems. Acta Horticulturae 1999;481: 213-220.

[8] Xu HL, Gauthier L, Gosselin A. Effects of fertigation management on growth and photosynthesis of tomato plants grown in peat, rockwool and NFT. Scientia Horticulturae 1995;6(1-2): 11-20.

[9] Massantini F, Favilli R, Magnani G, Oggiano N. Soilless culture-biotechnology for high quality vegetables. Soilless Culture 1988;4(2): 27-40.

[10] Asao T. Hydroponics - A Standard Methodology for Plant Biological Researches. Rijeka: InTech; 2012.

[11] Bilderback TE, Warren SL, Owen Jr. JS, Albano JP. Healthy substrates need physicals. HortTechnology 2005;15: 747-751.

[12] Mastouri F, Hassandokht MR, Padasht Dehkaei MN. The effect of application of agricultural waste compost on growing media and greenhouse lettuce yield. Acta Horticulturae 2005;697: 153-158.

[13] Olle M, Nagouajio M, Simos A. Vegetable quality and productivity as influenced by growing medium: A Review. Zemdirbyste Agriculture 2012;99(4): 399-408.

[14] Dorais M, Menard C, Begin E. Risk of phytotoxicity of sawdust substrate for greenhouse vegetables. Acta Horticulturae 2007;761: 589-594.

[15] Ehret DL, Helmer T. A new wood fibre substrate for hydroponic tomato and pepper crops. Canadian Journal of Plant Science 2009;89(6): 1127-1132.

[16] Mahamud S, Manisah MD. Preliminary studies on sago waste as growing medium for tomato. Acta Horticulturae 2007;742: 163-168.

[17] Nurznski J. The yield of greenhouse tomato grown in straw and rockwool. Folia Horticulture 2006;18(2): 17-23.

[18] Sevgican A. Protected vegetable growing. II. Ege Univeristy. Agriculture Faculty. Izmir, Turkey, 1999b; No. 526.

[19] Khalighi A, Padasht Dehkaei MN. The effect of media produced by tree bark, tea waste, rice hull and azolla as a substitute for peat, on growth and flowering marigold (*Tagetes patulta* L. Golden Boy). Iranian Journal of Agricultural Science 2000;31(3): 557-565.

[20] Urrestarazu M, Salas MC, Padilla MI, Moreno J, Elorrieta MA, Carasco GA. Evaluation of different composts from horticultural crop residue and their uses in greenhouse soilless cropping. Acta Horticulturae 2002;549: 147-152.

[21] Cotxarreara L, Trillas-Gay MI, Steinberg C, Alabouvette C. Use of sewage sludge compost and *Trichoderma asperellum* isolates to suppress *Fusarium* wilt of tomato. Soil Biology and Biochemistry 2002;34: 467-476.

[22] Coventry E, Noble R, Mead A, Whipps JM. Suppression of Allium white rot (*Sclerotium cepivorum*) in different soils using vegetables wastes. European Journal of Plant Pathology 2005;111: 101-112.

[23] Siddiqui Y, Sariah M, Ismail MR, Rahmani M, Ali A. Bio-efficiency of compost extracts on the wet rot incidence, morphological and physiological growth of okra (*Abelmoschus esculentus* [(L.) Moench.]). Scientia Horticulturae 2008;117: 9-14.

[24] Souleymane BK, Antoine D, Russell JT, Hani A, Tyler JA. Suppressive effect of non-aerated compost teas on foliar fungal pathogens of tomato. Biological Control 2010;52: 167-173.

[25] Cantliffe DJ, Funes J, Jovicich E, Paranjpe A, Rodriguez J, Shaw N. Media and containers for greenhouse soilless grown cucumbers, melons, peppers, and strawberries. Acta Horticulturae 2003;614: 199-203.

[26] Shaw NL, Cantliffe DJ, Funes J, Shine Ill. Successful Beit Alpha cucumber production in the greenhouse using pine bark as an alternative soilless media. HortTechnology 2004a;14: 289-294.

[27] Wilson GCS. Tomato production in bark substrates. Acta Horticulturae 1984;150: 271-276.

[28] Politycka B, Wojcik-Wojtkowiak D, Pudelski T. Phenolic compounds as a cause of phytotoxicity in greenhouse substrates repeatedly used in cucumber growing. Acta Horticulturae 1985;156: 89-94.

[29] Worrall RJ. The use of composted wood waste as a peat substitute. Acta Horticulturae 1978;82: 79-86.

[30] Gruda N, Prasad M, Maher MJ. Soilless Culture. In: Lal R. (ed.) Encyclopedia of Soil Science. Taylor & Francis (Marcel Dekker), Inc., New York; 2005.

[31] Gruda N, von Tucher S, Schenitzier WH. N-immobilization of wood fiber substrates in the production of tomato transplants (*Lycopersicon lycopersicum* (L). Karst. Ex Farw). Journal of Applied Botany 2000;74:32-37

[32] Garzo EC, Soria C, Gomez-Guillanmon ML, Fereres A. Feeding behavior of *Aphis gossypii* on resistant accessions of different melon genotypes *Cucumis melo*. Phytoparasitica 2002;30: 129-140.

[33] Beniot F, Ceusterman N. A decade of research on ecologically sound substrates. Acta Horticulturae 1995;408: 17-29.

[34] Tyson RV, Hochmuth RC, Lamb EE, Hochmuth GH, Sweat MS. A decade of change in Florida's greenhouse vegetable industry: 1991-2001. Proceedings of Florida State Horticultural Society, 2001;113: 280-282.

[35] Mumpton FA. Uses of natural zeolites in agriculture and industry. Proceedings of Natural Academy of Science, USA; 1999;96: 3463-3470.

[36] Allen ER, Ming DW. Recent progress in the use of natural zeolites in agronomy and horticulture. Nature of Zeolites 1995;93: 477-490.

[37] Dijedidi M, Grasopoulos D, Maloupa E. The effect of different substrates on the quality of f. Carmello tomatoes (*Lycopersicon escullentum* Mill.) grown under protection in a hydroponic system. Chaiers Options Mediterraneans 1997;31: 379-383.

[38] Issa M, Maloupa E, Gerasopoulos D. Effects of the substrate on yield and quality of two gerbera varieties grown under protection. Chaiers Options Mediterraneans 1997;31: 365-369.

[39] Kaveri Agri-Care. Processed coir pith (coco-peat) for export. Kaveri Agri-Care Pvt. Ltd., Bangalore, India; 26 January 2004.

[40] Donnan R. Hydroponics around the world. In: Practical Hydroponics & Greenhouses. July/August 1998.

[41] Seymour G. Review of commercial hydroponic crop production system. In: Commercial Hydroponics in Australia, A Guide for Growers, Pro-Set Pty Ltd, Hobart; 1993.

[42] Raviv M. The future of composts as ingredients of growing media. Acta Horticulturae 2011a;891: 19-32.

[43] Raviv M. Suppressing soil-borne disease of container-grown plants using composts. Acta Horticulturae 2011b;893: 169-181.

[44] Boodley JW, Sheldrake Jr. R. Cornel peat-lite mixes for commercial plant growing. International bulletin 43. New York State College of Agriculture and Life Science; 1977.

[45] Asaduzzaman M, Mondal FM, Ban T, Asao T. Selection of ideal succeeding crops after asparagus, taro, and beans replanting field in seedling growth bioassay. Allelopathy Journal 2013;32(2): 1-22.

[46] Hegde R, Miller DA. Allelopathy and autotoxicity in alfalfa: Characterization and effects of preceding crops and residue incorporation. Crop Science 1990;30: 1255-1259.

[47] Komada H. The occurrence, ecology of soil-borne diseases and their control. Takii Seed Co. Ltd. Kyoto, Japan 1988; p.1-3.

[48] Takahashi K. The replant failure of vegetables. Research Reports of National Research Institute of Vegetable and Tea Science, Japan 1994;pp. 87-99 (In Japanese).

[49] Sevgican A. Protected vegetable growing. I. Ege University. Agriculture Faculty, No. 528. Izmir, Turkey; 1999.

[50] Alan R. Characteristics of some of the growth media used in greenhouse. Turkey 5. Greenhouse Symposium, October 17-19, Izmir, Turkey; 1990: p.401-410.

[51] Gullino ML, Garibaldi A. Influence of soilless cultivation on soilborne diseases. Acta Horticulturae 1994;361: 341-353.

[52] Louvet J. The relationship between substrates and plant diseases. Acta Horticulturae 1982;126: 147-152.

[53] Riviere LM, Caron J. Research on substrates: State of the art and need for the coming 10 years. Acta Horticulturae 2001;548: 29-41.

[54] Jensen MH. Hydroponics. HortScience 1997;32: 1018-1021.

[55] Schwarz M. Soilless culture management. Springer-Verlag, New York, 1995.

[56] Reed DW. Water media and nutrition for greenhouse crops. In: Reed DW (ed.) Growing media: Types and physical/chemical properties. Ball Publishing, Batavia, Ill; 1996. p.93-122.

[57] Jensen MH. Greenhouse hydroponic industry status reports: Hydroponics worldwide. Acta Horticulturae 1999;481: 719-729.

[58] Hassall & Associates. Hydroponics as an agricultural production system: A Report for the Rural Industries Research and Development Corporation. RIRDC publication, no. 01/141; RIRDC project, no. HAS-9A. 2001. p.72.

[59] Dubik SP, Krizek DT, Stimart DP. Influence of root zone restriction on mineral element concentration, water potential, chlorophyll concentration, and partitioning of assimilate in spreading Euonymus (*E. kiautschovica Loes,* 'Sieboldina'). Journal of Plant Nutrition 1990; 677-699.

[60] Bar-Tal A. The significance of root size for plant nutrition in intensive horticulture. In: Renge Z. (ed.) Mineral Nutrition of Crops: Fundamental Mechanisms and Implications. New York: Haworth Press, Inc.; 1990. p.115-139.

[61] Varis S, Altay H. The most suitable and new method for soilless growing in Turkey: Perlite culture. First Perlite Symposium. Turkish Agriculture, Izmir, Turkey; 1992; p. 185.

[62] Abak K, Celikel G. Comparison of some Turkish originated organic and inorganic substrates for tomato soilless culture. Acta Horiculturae 1994; 366: 423-427.

[63] Alan RA, Zulkadir A, Padem H. The influence of growing media on growth, yield and quality of tomato grown under greenhouse conditions. Acta Horticulturae 1994;366: 429-436.

[64] Schnitzler WH, Gruda N. Quality issues of greenhouse production. Acta Horticulturae 2003;614: 663-674.

[65] Voogt W, de Visser PHE, van Winkel A, Cuijpers WJM, van de Burgt GJHM. Nutrient management in organic greenhouse production: Navigation between Constraints. Acta Horticulturae 2011;915: 75-82.

[66] Frezza D, Leon A, Logegaray V, Chiesa A, Desimone M, Diaz L. Soilless culture technology for high quality lettuce. Acta Horticulturae 2005;697:43-48.

[67] Lee SK, Kader AA. Preharvest and postharvest and postharvest factors influencing vitamin C content of horticultural crops. Postharvest Biology and Technology 2000;20(3):207-220.

[68] Singer SM, El-Tanahy AMM, El-Behairy UA, Abu El-Samad EH. Growth and productivity of cantaloupe plants grown under different soilless culture systems. Journal of Applied Sciences Research 2013;9(8): 5294-5302.

[69] Yang W, Chung S, Kim Y. Comparative studies on the physioecological and morphological adaptations of greenhouse tomato grown in aeroponics and nutrient film technique. III. Characteristics of physioecological adaptations. Journal of Korean Society of Horticultural Science 1990;31(3): 226-237.

[70] El-Behairy UA. The effect of levels of phosphorus and zinc in the nutrient solution on macro and micronutrients uptake and translocation in cucumber (*Cucumis sativus* L.) grown by the nutrient film technique. PhD thesis. London University 299P. UK, 1994.

[71] Gent PPN. Effect of nitrogen and potassium supply on yield and tissue composition of greenhouse tomato. Acta Horticulturae 2004;644: 369-375.

[72] Caretto S, Parente A, Serio F, Santamaria P. Influence of potassium and genotype on vitamin E content and reducing sugar of tomato fruits. Horticultural Science 2008;43: 2048-2051.

[73] Ramirez SLF, Diaz SFR, Muro EJ. Relation between soilless tomato quality and potassium concentration in nutritive solution. Acta Horticulturae 2012;947: 215-221.

[74] Benko B, Borosic J, Poljak M, Osvald J. Incidence of tomato blossom-end rot at different calcium levels. Acta Horticulturae 2012;927: 385-392.

[75] Nurzynski J. Effect of different fertilization levels on yielding of greenhouse tomato grown on sand, peat or rockwool growth media. Vegetable Crops Research Bulletin 2005;63: 101-107.

[76] Choi J, Ahn J, Ku J. Growth and nutrient uptake of tomato plug seedling influenced by elevated blendingrate of perlite in coir and peatmoss substrates. Horticulture Environment and Biotechnology 2007;48(5): 270-276.

[77] Miccolis V, Candido V, Lucarelli G, Castronuovo D. Cherry tomato yield on two different solid growing media. Acta Horticulturae 2007;761: 573-579.

[78] Komosa A, Kleiber T, Pirog J. Contents of macro- and micronutrients in root environment of greenhouse tomato grown in rockwool and wood fiber depending on nitrogen levels in nutrient solutions. Acta Scientiarum Polonorum Hortoroum Cultus 2010;9(3): 59-68.

[79] Rippy FMJ, Peet MM, Louws FJ, Nelson PV. Orr DB. Sorensen KA. Plant development and harvest yield of greenhouse tomatoes in six organic growing systems. HortScience 2004;39(2): 223-229.

[80] Kowalczyk K, Gajc-Wolska J, Marcinkoswska M. The influence of growing medium and harvest time on the biological value of cherry fruit and standard tomato cultivars. Vegetable Crop Research Bulletin 2011a;74: 51-59.

[81] Kowalczyk K, Gajc-Wolska J, Radzanowska J, Marcinkoswska M. Assessment of chemical composition and sensory quality of tomato fruit depending on cultivars and growing conditions. Acta Scientiarum Polonorum Hortorum Cultus 2011b;10(4): 133-140.

[82] Nurzynski J. Yield and quality of greenhouse tomato fruit grown in rape straw substrates. Acta Scientiarum Polonorum Hortorum Cultus 2013;12(1): 3-11.

[83] Permuzic Z, Bargiela M, Garci A, Rendina A. Calsium, iron, potassium, phosphorous and vitamin C content of organic and hydroponic tomatoes. Hort Science 1998;33(2): 255-257.

[84] Inden H, Torres A. Comparison of four substrates on the growth and quality of tomatoes. Acta Horticulturae 2004;644: 205-210.

[85] Djedidi M, Gerasopoulos D, Maloupa E. The effect of different substrates on the quality of F. Carmello tomatoes grown under protection in a hydroponics system. Cahiers Option Mediterraneans's 2001;p.31.

[86] Fernández-Trujillo JP, Sánchez C, Obando J, Gómez MD, Mercader JM. Quality of greenhouse fruit grown on perlite substrate or nutrient film technique. Acta Horticulturae 2004;633: 229-236.

[87] Tzortzakis NG, Economakis CD. Impact of the substrate medium on tomato yield and fruit quality in soilless cultivation. Horticultural Science (Prague) 2008;35: 83-89.

[88] Dysko J, Kowalczyk W, Kaniszewski S. The influence of ph of nutrient solution on yield and nutritional status of tomato plants grown in soilless culture system. Vegetable Crop Research Bulletin 2009;70: 59-69.

[89] Thybo AK, Bechmann IE, Brandt K. Integration of sensory and objective measurements of tomato quality: quantitative assessment of the effect of harvest date as compared with growth medium (soil *versus* rockwool), electrical conductivity, variety and maturity. Journal of the Science of Food and Agriculture 2005;85: 2289-2296.

[90] Tuzel JH, Tuzel Y, Sul A, Elterz RZ. Effect of EC level on the nutrient solution on yield and fruit quality of tomato. Acta Horticulturae 2001;559: 587-592.

[91] Dorais M, Papadopulos AP, Gosselin A. Influence of electric conductivity management on greenhouse tomato yield and fruit quality. Agronomie 2001;21: 367-383.

[92] Adams P, Ho LC. Effects of constant and fluctuating salinity on the yield, quality and calcium status of tomatoes. Journal of Horticultural Science 1989;64: 725-732.

[93] Petersen K, Willumsen J, Kaack K. Composition and taste of tomatoes as affected by increased salinity and different salinity sources. Journal of Horticultural Science & Biotechnology 1998;73: 205-215.

[94] Sato S, Sakaguchi S, Furukawa H, Ikeda H. Effects of NaCl application to hydroponic nutrient solution on fruit characteristics of tomato (*Lycopersicon esculentum* Mill.). Scientia Horticulturae 2006;109: 248-253.

[95] Adams P. Effects of increasing the salinity of the nutrient solution with major nutrients or sodium chloride on the yield, quality and composition of tomatoes grown in rockwool. Journal of Horticultural Science 1991;66: 201-207.

[96] Serio F, De Gara L, Caretto S, Leo L, Santamaria P. Influence of an increased NaCl concentration on yield and quality of cherry tomato grown in posidonia (*Posidonia oceanica* (L.) Delile). Journal of the Science of Food and Agriculture 2004;84: 1885-1890.

[97] Tzortzakis NG, Economakis CD. Shredded maize stems as an alternative substrate medium - Effect on growth, flowering and yield of tomato in soilless culture. Journal of Vegetation Science 2005;11: 57-70.

[98] Tzortzakis NG, Economakis CD. Impact of the substrate medium on tomato yield and fruit quality in soilless cultivation. Horticultural Science (Prague) 2008;35: 83-89.

[99] Vesseur WP. Tomato tasting and consumer attitude. Acta Horticulturae 1990; 259: 83-89.

[100] Davies J., Hobson G. The constituents of tomato fruit-the influence of environment, nutrition and genotype. In: McGlasoon W. CRC Critical Reviews in Food Science and Nutrition. Taylor and Francis 181;15: 205-281.

[101] Fandi M, Al-Muhtaseb JA, Hussein MA. Yield and fruit quality of tomato as affected by the substrate in an open soilless culture. Jordan Journal of Agricultural Sciences, 2008;4(1): 65-72.

[102] Siomos AS, Beis G, Papadopoulou PP, Barbayiannis N. Qulaity and composition of lettuce (cv. Plenty) grown in soil and soilless culture. Acta Horticulturae 2001;548: 445-450.

[103] Luitel B, Adhikari PB, Yoon CS, Kang WH. Yield and fruit quality of tomato (*Lycopersicon esculentum* Mill.) cultivars established at different planting bed size and growing substrate. Horticulture Environment and Biotechnology 2012;53(2): 102-107.

[104] Mazur KZ, Gajewski M, Metera AM, Wtulich JA, Marcinkowska MM. Effect of growing medium and harvest term on yield and several quality traits of two cultivars of 'Cherry' tomatoes. Notulae Botanicae Horti Agrobotanici 2012;40(2): 197-202.

[105] Islam MS, Matsui T, Yoshida Y. Carbohydrate content and the activities of sucrose synthase, sucrose phophate synthase and acid invertase in different tomato cultivars during fruit development. Scientia Horticulturae 1996;65: 125-136.

[106] Padem H, Alan R. The Effects of some substrate on yield and chemical composition of peppers under greenhouse conditions. Acta Horticulturae 1994;366: 445-451.

[107] Schnitzler WH, Sharma AK, Grud NS, Heuberger HT. A low-tech hydroponics system for bell pepper (*Capsicum annuum* L.). Acta Horticulturae 2004;644: 47-52.

[108] Gungor F, Yildirim E. Effect of different growing media on quality, growth and yield of pepper (*Capsicum annuum* L.) under greenhouse conditions. Pakistan Journal of Botany 2013;45(5): 1605-1608.

[109] Aydemir O, Ince F. Plant nutrition. Dicle University, Diyarbakir, Turkey, Publication no. 2; p.653

[110] Majdi Y, Ahmadizadeh M, Ebrahimi R. Effect of different substrates on growth indices and yield of green peppers at hydroponic cultivate. Current Research Journal of Biological Sciences 2012;4(4): 496-499.

[111] Maaswinkel R, Gunadi N, Lembang I. Improving sweet pepper yields in Indonesia. Fruit and Vegetable Technology 2009;9(4): 14-15.

[112] Albaho M, Bhat N, Abo H, Tomas B. Effect of three substrates on growth and yield of two cultivars of *Capsicum annuum*. European Journal of Scientific Research 2009;28: 233-227.

[113] Lopez- Madina J, Perablo A, Flores F. Closed soilless system growing: A sustainable solution to strawberry crop in Huelva (Spain). Acta Horticulturae 2008;649: 213-215.

[114] Ercisli S, Sahin U, Esitken A, Anapali O. Effects of some growing media on the growth of strawberry cvs. 'Camarosa' and 'Fern'. Acta Agrobotanica 2005;58: 185-191.

[115] Jafarnia S, Khosrowshahi S, Hatamzadeh A, Tehranifar A. Effect of substrate and variety on some important quality and quality characteristics of strawberry production in vertical hydroponic system. Advances in Environment and Biology 2010;4(3): 360-363.

[116] Caso C, Chang M, Rodriguez-Delfin A. Effect of growing media on the strawberry production in column system. Acta Horticulturae 2009;843: 373-380.

[117] Schmitzer V, Stampar F. Changes in anthocyanin and selected phenolics in Korcrisett rose flowers due to substrate pH and foliar application of sucrose. Acta Horticulturae 2009;870: 89-96.

[118] Lopes da Silva F, Escribano-Bailon MT, Joaquin Perez Alonso J, Rivas-Gonzalo JC, Santos-Buelga C. Anthocyanin pigments in strawberry. LWT- Food Science and Technology 2007;40: 374-382.

[119] Ameri A, Tehranifar A, Davarynejad GH, Shoor M. The effect of substrate and cultivar in quality of strawberry. Journal of Biological and Environmental Science 2012; 6(17): 181-188.

[120] Tulipani S, Mezzetti B, Capocasa F, Bompare S, Beekwilder J. Antioxidants, phenolic compounds, and nutritional quality of different strawberry genotypes. Journal of Agriculture and Food Chemical 2008;56: 696-704.

[121] Chellemi DO, Rosskopf EN. Yield potential and soil quality under alternative crop production practices for fresh market pepper. Renewable Agriculture and Food Systems 2004;19: 168-175.

[122] Poudel DD, Horwath WR, Lanini WT, Temple SR, Bruggen AHC. Comparison of soil N availability and leaching potential, crop yields and weeds in organic, low-input and conventional farming systems in northern California. Agriculture Ecosystem and Environment 2002;90: 125-137.

[123] Curuk S, Sermenli T, Mavi K, Evrendilek F. Yield and fruit quality of watermelon (*Citrullus lanatus* (Thumb.) Matsum. & Nakai.) and melon (*Cucumis melo* L.) under protected organic and conventional farming systems in a Mediterranean region of Turkey. Biological Agriculture and Horticulture 2004;22: 171-183.

[124] Flores P, Hellin P, Lacasa A, Lopez A, Fenoll J. Pepper mineral composition and sensory attributes as affected by agricultural management. Journal of the Science of Food and Agriculture 2009;89: 2364-2371.

[125] Fukuda N, Anami Y. Substrate and nutrient level: Effects on the growth and yield of melon '*Cucumis melo*' in soilless culture. Acta Horticulturae 2002;588: 111-117.

[126] Guler HG, Olympios C, Gerasopoulos D. The effect of the substrate on the fruit quality of hydroponically grown melons *Cucumis melo* L. Acta Horticulturae 1995;379: 261-266.

[127] Bohme M. Evaluation of organic, synthetic and mineral substrates for hydroponically grown cucumbers. Acta Horticulturae 1995;401: 209-217.

[128] Schwarz M. Soilless culture management. Springer-Verlag, New York, 1995.

[129] Rodriguez JC, Cantliffe DJ, Shaw NL, Karchi Z. Soilless media and containers for greenhouse production of 'Galia' type musk melon. HortScience 2006;41: 1200-1205.

[130] Zulkarami B, Ashrafuzzaman M, Mohd Razi I. Morpho-physiological growth, yield and fruit quality of rock melon as affected by growing media and electrical conductivity. Journal of Food, Agriculture & Environment 2010;8(1): 249-252.

[131] Yetisir H, Sari N, Aktas H, Karaman C, Abak K. Effect of different substrates on plant growth yield and quality of watermelon grown in soilless culture. Agriculture and Environmental Science 2006;1(2): 113-118.

[132] Jankauskiene J, Brazaityte A. The influence of various substratum on the quality of cucumber seedlings and photosynthesis parameters. Scientific Works of the Lithuani-

an Institute of Horticulture and Lithuanian University of Agriculture. Sodininkyste Ir Darzininkyste 2008;27(2): 285-294.

[133] Alifar N, Mohammadi Ghehsareh A, Honarjoo N. The effect of growth media on cucumber yield and its uptake of some nutrient elements in soilless culture. Journal of Science and Technology of Greenhouse Culture. Isfahan University of Technology 2010;1: 19-25.

[134] Ghehsareh AM, Hematian M, Kalbasi M. Comparison of date-palm wastes and perlite as culture substrates on growing indices in greenhouse cucumber. International Journal of Recycling of Organic Waste in Agriculture 2012;1(5): 1-4.

[135] Asaduzzaman M, Kobayashi Y, Mondal MF, Ban T, Matsubara H, Adachi F, Asao T. Growing carrots hydroponically using perlite substrates. Scientia Horticulturae 2013;159: 113-121.

[136] Willis CR. Hydroponics-the Solution to the Food Problem. 2009. http://www.androidpubs.com/Chap02.htm,10/07/09].

[137] Ahmad Z, Khan MA, Qasim M, Zafar MS, Ahmad R. Substrate effects on growth, yield and quality of *Rosa hybrida* L. Pakistan Journal of Botany 2012;44(1): 177-185.

[138] Fascella G, Zizzo GV. Effect of growing media on yield and quality of soilless cultivated rose. Acta Horticulturae 2005;697: 133-138.

[139] Wahome PK, Oseni TO, Masarirambi MT, Shongwe VD. Effects of different hydroponics systems and growing media on the vegetative growth, yield and cut flower quality of gypsophila (*Gypsophila paniculata* L.). World Journal of Agricultural Sciences 2011;7(6): 692-698.

[140] Ryota T, Kazunori M, Musao T, Kazuyoshi S. Studies on the hydroponics of oriental hybrid lily. Journal of the Niigata Agricultural Research Institute 2002;5: 65-74.

[141] Hsu JH, Lin YH, Lin RS. Effect of cultural medium and hydroponics culture on growth and flower quality of *Oncidium* grower Ramsey. Acta Horticulturae 2007;761: 489-493.

[142] Raja Harun RM, Hall DA, Szmidt RAK, Hitchon GM. Melon cultivation in organic and inorganic substrates. Acta Horticulturae 1991;294: 105-108.

[143] Nurzynski J, Michalojc Z, Jarosz Z. Mineral nutrient concentration in potting media (rockwool, peat, sand) and growth of tomato. Vegetable Crops Research Bulletin 2001;55: 45-48.

[144] Elia A, Frezza D, Fraschina A, Trinchero G, Moccia S, Leon A. Preharvest factors and fresh cut vegetables quality. Acta Horticulturae 1999;481: 267-271.

[145] Gruda N, Schnitzler WH. Suitability of wood fiber substrates for production of vegetable transplants. I. Physical properties of wood fiber substrates. Scientia Horticulturae 2004a;100: 309-322.

[146] Gruda N, Schnitzler WH. Suitability of wood fiber substrates for production of vegetable transplants. II. The effect of wood fiber substrates and their volume weights on the growth of tomato transplants. Scientia Horticulturae 2004b;100: 333-340.

[147] Gruda N, Schnitzler WH. The effect of water supply on bio-morphological and plant-physiological parameters of tomato transplants cultivated in wood fiber substrate. Journal of Applied Botany 2000a;74: 233-239.

[148] Gruda N, Schnitzler WH. The effect of water supply of seedlings cultivated in different substrates and raising systems on the bio-morphological and plant-physiological parameters of lettuce. Journal of Applied Botany 2000b;74: 240-247.

[149] Gruda N, Schnitzler WH. Wood fiber substrates as a peat alternative. European Journal of Wood and Wood Products 2006;64: 347-350.

[150] Luthria DL, Mukhopadhyaya S, Krizek DT. Content of total phenolics and phenolic acids in tomato (*Lycopersicon esculentum* Mill.) fruits as influenced by cultivar and solar UV radiation. Journal of Food Composition and Analysis 2006;19: 771-777.

[151] Martinez-Valverde I, Periago MJ, Provan G, Chesson A. Phenolic compounds, lycopene and antioxidant activity in commercial varieties of tomato (*Lycopersicon esculentum* L.). Journal of the Science of Food and Agriculture 2002;82: 323-330.

[152] Gruda N. Do soilless culture systems have influence on product quality of vegetables? Journal of Applied Botany and Food Quality 2009;82: 141-147.

[153] Santamaria P, Valenzano V. La qualita degli ortaggi allevati senza suolo. Italus Hortus 2001;8: 31-38.

[154] Cefola M, Pace B, Buttaro D, Santamaria P, Serio F. Postharvest evaluation of soilless-grown table grape during storage in modified atmosphere. Journal of Science of Food and Agriculture 2011;91: 2153-2159.

[155] Pedrero Z, Madrid Y, Camaraj C. Selenium species bioaccessibility in enriched radish (*Raphanus sativus*): a potential dietary source of selenium. Journal of Agriculture and Food Chemistry 2006;54: 2412-2417.

[156] Inoue K, Kondo S, Adachi A, Yokota H. Production of iron enriched vegetables: effect of feeding time on the rate of increas in foliar iron content and foliar injury. The Journal of Horticultural Science and Biotechnology 2000;75: 209-213.

[157] Palaniswamy UR, McAvoy RJ, Bible BB. Omega-3-fatty acid concentration in *Portulaca oleracea* is altered by nitrogen source in hydroponic solution. Journal of the American Society for Horticultural Science 2000;125: 190-194.

[158] Santamaria P, Elia A, Gonnella M, Parente A, Serio F. Ways of reducing rocket salad nitrate content. Acta Horticulturae 2001;548: 529-537.

[159] Asao T, Asaduzzaman M, Mondal MF, Tokura M, Adachi F, Ueno M, Kawaguchi M, Yano S, Ban T. Impact of reduced potassium nitrate concentrations in nutrient solu-

tion on the growth, yield and fruit quality of melon in hydroponics. Scientia Horticulturae 2013;164: 221-231.

[160] Brandt K, Molgaard JP. Organic agriculture: Does it enhance or reduce the nutritional value of plant foods? Journal of Science of the Food and Agriculture 2001;81: 924-931.

[161] Chessy AW, Bui L, Renaud ENC, van Horn M, Mitchell AE. Three years comparison of the content of antioxidant microconstituents and several quality characteristics in organic and conventionally managed tomato and bell peppers. Journal Agriculture and Food Chemistry 2006;54: 8244-8252.

[162] Heeb A, Lundegardh B, Ericsson T, Savage GP. Nitrogen affects yield and taste of tomatoes. Journal of the Science of Food and Agriculture 2005;85: 1405-1414.

[163] Caris-Veyrat C, Amiot MJ, Tyssandier V, Grasselly D, Buret M, Mikolajczak M, et al. Influence of organic versus conventional agriculture practice on the antioxidant microconstituents content of tomatoes and derived purees; consequences on antioxidant plasma status in humans. Journal of Agriculture and Food Chemistry 2004;52: 6503-6509.

[164] Pieper JR, Barrett DM. Effect of organic and conventional production systems on quality and nutritional parameters of processing tomatoes. Journal of the Science of Food and Agriculture 2009;89: 177-194.

[165] Hallmann E. The estimation of crop and quality fruits of selected types of tomato cultivated on mineral rockwool. PhD thesis, Department of Horticulture, WULS-SGGW; 2003.

[166] Riahi A, Hdiner CH, Sanaa M, Tarchoun N, Kheder MB, Guezal N. Effect of conventional and organic production system on the yield and quality of field tomato cultivars grown in Tunisia. Journal of the Science of Food and Agriculture 2009;89: 2275-2282.

[167] Worthington V. Nutritional quality of organic versus conventional fruits, vegetables, and grains. Journal of Alternative and Complementary Medicine 2001;7: 11-173.

[168] Hallmann E. The influence of organic and conventional cultivation systems on the nutritional value and content of bioactive compounds in selected tomato types. Journal of Science of the Food and Agriculture 2012;92: 2840-2848.

[169] Auerswald H, Schwarz D, Kornelson C, Krumbein A, Bruckner B. Sensory analysis, sugar and acid content of tomato at different EC values of the nutrient solution. Scientia Horticulturae 1999;82(3-4): 227-242.

[170] Lacatus V, Botez C, Chelu M, Mirghis R, Voican V. The influence of organic and mineral fertilizers on tomato quality for processing. Acta Horticulturae 1994;276: 329-332.

[171] Naidu KA. Vitamin C in human health and disease is still a mystery? An overview. Nutrition Journal 2003;2: 7.

[172] Sima R, Maniutiu D, Apahidean AS, Apahidean M, Lazar V, Muresan C. The influence of fertilization on greenhouse tomatoes cultivated in peat bag system. Bulletin UASVM Horticulture 2009;66(1): 455-460.

[173] Gannella M, Serio F, Conversa P, Santamarina P. Yield and qualtiy of lettuce grown in floating system using different sowing density and pant spatial arrangements. Acta Horticulturae 2003;614: 687-692.

[174] Siomos AS, Beis G, Papadopoulou PP, Barbayiannis N. Quality and composition of lettuce (cv. "Plenty") grown in soil and soilless culture. Acta Horticulturae 2001;548: 445-449.

[175] Omran MS, Waly TM, Abd Elnaim EM, El Nashar BB. Effect of sewage irrigation on yield tree components and heavy metal accumulation in Navel orange trees. Biological Wastes 1988;23: 17-24.

[176] Yadav RK, Goyal B, Sharma RK, Dubey SK, Minhas PS. Post-irrigation impact of domestic sewage effluent on composition of soils, crops and ground water - A case study. Environment International 2002;28: 481-486.

[177] Banin AJ, Naverot YN, Yoles D. Accumulation of heavy metal in arid zone soils irrigated with treated sewage effluent and their uptake by Rhodes grass. Journal of Environmental Quality 1981;10(4): 536-540.

[178] Mapanda F, Mangwayana EN, Nyamangara J, Giller KE. The effect of long term irrigation using wastewater on heavy metal contents of soils under vegetables in Harare, Zimbabwe. Agriculture Ecosystem and Environment 2005;107: 151-165.

[179] Mocquot B, Vangronsveld J, Clijsters H, Mench M. Copper toxicity in young maize (*Zea mays* L.) plants: Effects on growth, mineral and chlorophyll contents, and enzyme activities. Plant and Soil 1996;182: 287-300.

[180] Krishnamurti GSR, Naidu R. Speciation and phytoavailability of cadmium in selected surface soil of South Australian. Australian Journal of Soil Research 2000;38: 991-1004.

[181] Cheng S. Effects of heavy metals on plants and resistance mechanisms. A state-of-the-art report with special reference to literature published in Chinese journal. Environmental Science and Pollution Research 2003;10: 256-264.

[182] Guo GL, Zhou QX, Koval PV, Belogolova GA. Speciation and distribution of Cd, Pb, Cu and Zn in contaminated Phaeozem in north-east China using single and sequential extraction procedures. Australian Journal of Soil Research 2006;44: 135-142.

[183] Mclaughlin WI, Parker RR, Clarke JM. Metals and micronutrients-Food Safety Issues. Field Crop Research 1999;60: 143-163.

[184] Chojnacki K, Chojnacki A, Gorecka H, Gorecki H. Bioavailability of heavy metals from polluted soils to plants. Science of the Total Environment 2005;337: 175-182.

[185] Emongor VE, Ramoleman, GM. Treated sewage effluents (water) potential to be used for horticultural production in Botswana. Physics and Chemistry of the Earth 2004;29: 1101-1108.

[186] Kalavrouziotis IK, Robolas PK, Koukoulakis PH, Papadopoulos AH. Effects of municipal reclaimed water on the macro-, micronutrients of soil and *Brassica oleracea* var. Italica, and *Brassica oleracea* var. gemmifera. Journal of Agricultural Water Management 2008;95(4): 419-426.

[187] Emongor VE, Macheng BJ, Kefilwe S. Effects of secondary sewage effluent on the growth, development, fruit yield, and quality of tomato (*Lycopersicum lecopersicum* (L.) Karten). International Symposium on Vegetable Production, Quality and Process Standardization in China: a Worldwide Perspective. Acta Horticulturae 2012;944: 29-40.

[188] Asano T, Pettygrove GS. Using reclaimed municipal wastewater for irrigation. California Agriculture 1987;41: 16-18.

[189] Hussain I, Raschid L, Hanjra MA, Marikar F, van der Hoek W. Wastewater use in agriculture: Review of impacts and methodological issues in valuing impacts. Working Paper 37. International Water Management Institute, Colombo, Sri Lanka, 2002.

[190] Sanderson KC. Introduction to the workshop on wastewater utilization in horticultue. HortScience 1986;21: 23-24.

[191] Bozkurt MA, Yaliga T. The effects of waste water sludge application on the yield, growth, nutrition and heavy metal accumulation in apple trees growing in dry conditions. Turkish Journal of Agriculture and Forestry 2003;27: 285-292.

[192] Kiziloglu FM, Turan M, Sahin U, Kuslu Y, Dursun A. Effects of untreated and treated wastewater irrigation on some chemical properties of cauliflower (*Brassica oleracea* L. var. botrytis) and red cabbage (*Brassica oleracea* L. var. rubra) grown on calcareous soil in Turkey. Agriculture Water Management 2008;95: 716-724.

[193] Wattanapreechanon K, Wattanapreechanon E. Development of soilless culture for crop production at Chitralada Palace. Towards the year 2000: Technology for Rural Development. Proceedings of the International Conference. Chulalongkorn University, Bangkok, Thailand, August 25-26, 1997.

[194] Paul C. Health and hydroponics. Practical Hydroponics & Greenhouse 2000;53(4): 28-30.

[195] Jiang WJ, Yu HJ. Current situation and perspective of protected horticulture in China. International Workshop: The Production in the Greenhouse after the Era of the Methyl Bromide. April 1-3, 2004, Comiso, Italy.

[196] Ratanakosl P. Soilless culture not complex as we think. Kasikorn 1997;6(6): 587-592.

[197] Hanna HY. Influence of cultivar, growing media, and cluster pruning on greenhouse tomato yield and fruit quality. HortTechnology 2009;19(2): 395-399.

[198] Hanna HY, Properly recycled perlite saves money, does not reduce greenhouse tomato yield, and can be reused for many years. HortTechnology 2005;15: 342-345.

[199] Hanna HY, A stir and distinct technique to recycle perlite for cost effective greenhouse tomato production. Journal of Vegetation Science 2006;12(1): 51-63.

[200] Hanna HY, Smith DT. Recycling perlite for more profit in greenhouse tomatoes. Louisiana Agriculture 2002;45: 9.

[201] Chretien SA, Gosselin A, Dorais M. High electrical conductivity and radiation-based water management improve fruit quality of greenhouse tomatoes grown in rockwool. HortScience 2000;35: 627-631.

[202] del Amor FM, Martinez V, Cerda A. Salt tolerance of tomato plants as affected by stage of plant development. HortScience 2001;36: 1260-1263.

[203] Schwarz D. Concentration and composition of nutrient solution affect root formation of young tomato. Acta Horticulturae 2003;609: 103-108.

[204] Papadopoulos AP. Growing greenhouse tomatoes in soil and in soilless media. Publication 1865/E. Communications Branch, Agriculture and Agri-food Canada, Canada; 1991.

[205] Straver WA. Inert growing media for greenhouse tomatoes. In: Proceedings of Greenhouse Tomato Seminar. ASHS Seminar Series, American Society for Horticultural Science Press, Alexandria, VA, USA; 1995.

[206] Benoit F Ceustermans N. Consequences of Closed soilless growing systems for the recirculating nutrient solution and the production techniques. Acta Horticulturae 2004;633: 331-340.

[207] Van Noordwijk M. Synchronization of supply and demand is necessary to increase efficiency of nutrient use in soilless horticulture. In: van Beusichem ML (ed.). Plant Nutrition-Physiology and Applications. Kluwer Academic Publishers. 1990; p. 525-531.

[208] Incroccci L, Malorgio F, Della Bartola A, Pardossi A. The influence of drip irrigation or subirrigation on tomato grown in closed loop substrate culture with saline water. Scientia Horticulturae 2006;107: 365-372.

[209] Rumia TW, van Os EA, Bollen GJ. Disinfection of drain water from soilless culture by heat treatment. Netherlands Journal of Agricultural Science 1988;36: 231-238.

[210] Voogt W. Meststofverbruik: Realizatie en Normerbruik. Applied Plant Research, Naaldwijk Internal Report, 2003;p.15.

[211] Sonneveld C. Mineralenbalansen Bij Kasteelten. Meststoffen, 1993;p.44-49.

[212] Sonneveld C. The salt tolerance of greenhouse crops. Netherlands Journal of Agricultural Science 1988;36: 63-73.

[213] Sonneveld C. van den Bos AL. Effects of nutrient levels on growth and quality of radish (*Raphanus sativus* L.) grown on different substrates. Journal of Plant Nutrition 1995;18: 501-513.

[214] Ruijs MNA. Van Grondteelt Naar Gesloten Teelt. Groenten en Fruit Vakdeel Glasgroenten, 1992;2: 64-65.

[215] Abd-Elmoniem EM, El-Behairy UA. Effect of reusing perlite and pumice as a substrate on yield and mineral composition of strawberry plants. Egyptian Journal of Horticulture 2004;31: 13-21.

[216] Rea E, Salerno A, Pierandrei F. Effect of substrate and nutrient solution reuse on ranunculus and anemone plant production in a closed soilless system. Acta Horticulturae 2008;779: 542-546.

[217] Celikel G, Caglar G. The effects of re-using different substrates on the yield and earliness of cucumber on autumn growing period. Acta Horticulturae 1999;492: 259-264.

[218] Giuffrida F, Leonardi C, Marfa O. Substrate reuse in tomato soilless cultivation. Acta Horticulturae 2007;801: 1577-1582.

[219] Acuna R, Bonachela S, Magan JJ. Response of a sweet pepper crop grown in new and two-year-old reused rockwool slabs in greenhouse on the Mediterranean coast of south-east Spain. Acta Horticulturae 2005;697: 189-194.

[220] Fernandes C, Cora JE, Braz LT. Effects of the reutilization of substrates on its physical properties on growing cherry tomato. Horticulturae Brasileira 2006;24: 94-98.

[221] Urrestarazu M, Mazuela PC, Martinez GA. Effect of substrate reutilization on yield and properties of melon and tomato crops. Journal of Plant and Nutrition 2008;31: 2031-2043.

[222] Montero JI, Anton A, Munoz P, Lorenzo P. Transpiration from geranium grown under high temperatures and low humidifies in greenhouses. Agricultural and Forest Meteorology 2009;107: 323-332.

[223] Benoit F, Cuestermans N. A decade of research on polyurethane foam (PUR) substrate. Plasticulture 1994;104: 47-53.

[224] Hardgrave M. An evaluation of polyurethane foam as a reusable substrate for hydroponic cucumber production. Acta Horticulturae 1995;401: 201-208.

[225] Wilson GCS. The effect of various treatments on the yield of tomatoes in re-used perlite. Acta Horticulturae 1988;221: 373-382.

[226] Kang TM, Jung HB. Effect of rock wool classification on the growth and yield in long term culture of tomatoes. Journal Korean Society of Horticultural Science 1995;13: 352-353.

[227] Beavre OA. Base fertilizer and re-using of peat bag for tomato. Acta Horticulturae 1980;99: 11-16.

[228] Kampf AN, Jung M. The reuse of carbonized rice hulls as a horticultural substrate. Acta Horticulturae 1991;294: 271-283.

[229] Aendekerk TGL. Decomposition of peat substrate in relation to physical properties and growth of Chamaecyparis. Acta Horticulturae 1997;450: 191-198.

[230] Aendekerk TGL. Decomposition of peat substrate in relation to physical properties and growth of Skimmia. Acta Horticulturae 2001;548: 261-268.

[231] Prasad M, O'Shea J. Relative breakdown of peat and non-peat growing media. Acta Horticulturae 1999;401: 473-486.

[232] Pardossi A, Carmassi G, Diara C, Incrocci L, Maggini R, Massa D. Fertigation and substrate management in closed soilless culture. Dipartimento di Biologia delle Piante Agrarie, University of Pisa, Italy; 2011, p.63.

[233] Djedidi M, Geraspoulos D, Maloupa E. The effect of different substrates on the quality of F. Carello tomato (*Lycopersicon esculentum* Mill.) grown under protection in a hydroponic system. Cahier Option Mediterranneans 1999;31: 379-383.

[234] Hassanpouraghdam MB, Tabatabaei SJ, Nazemiyeh L, Vojodi L, Aazami MA. Essential oil composition of hydroponically grown *Chrysanthemum balsamita* L. Journal of Essential Oil Bearing Plants 2008;11: 649-654.

[235] Hassanpouraghdam MB. Flowerheads volatile oil composition of soilless culture grown *Chrysanthemum balsamita* L. Natural Product Research 2009;23: 672-677.

[236] Mairapetyan SK. Aromatic plant culture in open air hydroponics. Acta Horticulturae 1999;503: 33-42.

[237] Manukyan AE, Heuberger HT, Schnitzler WH. Yield and quality of some herbs of the Laminaceae family under soilless greenhouse production. Journal of Applied Botany and Food Quality 2004;78: 193-199.

[238] Dorais M, Papadopoulos AP, Luo X, Leonhart S, Gosselin A, Pedneault K, Angers P, Gaudreau L. Soilless greenhouse production of medicinal plants in North Eastern Canada. Acta Horticulturae 2001;554: 297-304.

[239] Hassanpouraghdam MB, Tabatabaei SJ, Nazemiyeh H, Aflatuni A. N and K nutrition levels affect growth and essential oil content of Costmary (*Tanacetum balsamita* L.). Journal of Food Agriculture and Environment 2008;6: 150-154.

[240] Suh E, Park K, Park K. Effect of different concentrations of nutrient solutions on the growth, yield and quality of basil. Acta Horticulturae 1999;502: 56-61.

[241] Udagawa Y. Some responses of Dill (*Anethum graveolens*), grown in hydroponic, to the concentration of nutrient solution. Acta Horticulturae 1995;396: 203-210.

[242] Hassanpouraghdam MB, Tabatabaei SJ, Aazami MA, Shekari F. Soilless culture production of Alecost [*Chrysanthemum balsamita* (L) Baill.]: A Preliminary Study. Romanian Biotechnological Letters 2010;15(5): 5530-5536.

[243] Haddad M. Effect of three substrates on growth, yield and quality of tomato by the use of geothermal water in the south of Tunisia. Journal of Food Agriculture and Environment 2007;5(2): 175-178.

[244] Olympios CM. Soilless media under protected cultivation rockwool, peat, perlite and other substrates. Acta Horticulturae 1995;401: 443-451.

Chapter 2

Design and Preparation of Nutrient Solution Used for Soilless Culture of Horticultural Crops

J.A. Olfati

Additional information is available at the end of the chapter

http://dx.doi.org/10.5772/59478

1. Introduction

Hydroponic systems include all systems that deliver the nutrients in a liquid form, with or without an aggregate medium to anchor the plant roots. Hydroponic systems in controlled environments can produce high quality plant free from accidental adulteration by weeds, soil or environmental toxins such as heavy metals in soils. In some species it may be possible to optimize for higher yields of target organs [1].

2. Mineral nutrition required for plants growing in soilless hydroponics

2.1. Nitrogen (N)

Nitrogen is the most frequently limiting nutrient. Within the plant, nitrogen serves in the same ways it does in other organisms—as a component of amino acids and nucleic acids. Nitrogen also plays a critical role in the structure of chlorophyll, the primary light harvesting compound of photosynthesis. This, along with its structural role in amino acids, explains why plants require large amounts of nitrogen, and thus why it is often the limiting nutrient for plant growth. The largest natural source of nitrogen is the Earth's atmosphere, which is roughly 78% gaseous nitrogen, an inert and essentially biologically unavailable form of the element. Its biological unavailability is because the two nitrogen atoms form an extremely stable bond, which is not easily broken. Apart from human industrial processes that fix nitrogen gas to solid or liquid forms, the primary means of nitrogen fixation are through the high temperature and energy of lightning strikes and biological nitrogen fixation by bacteria. These processes produce nitrogen in three main forms, each of which is available to plants: nitrate, nitrite, and ammonium. Nitrogen deficiency is commonly revealed by chlorosis. In the case of nitrogen-deficient chlorosis, the effects are first seen in the more mature leaves and tissues. The plant will preferentially export nitrogen to actively growing tissues, leaving the more mature parts of the plant to show signs of deficiency first. Nitrogen deficiency affects not only the leaves of the plant, but all living cells that have high nitrogen demands for amino and nucleic acids, reducing overall productivity and plant vigor. Generally, nitrogen-deficient plants also exhibit the spindly growth of an etiolated habit [2].

2.2. Phosphorus (P)

Phosphorus is frequently a limiting nutrient, particularly in tropical regions, where the soil chemistry differs from temperate soils, or in highly weathered soils, where phosphorus has long since leached away. Phosphorus is one of the three main elements in commercial lawn fertilizers, though there is mounting evidence that many lawns and green areas already have ample phosphorus, and thus it is being phased out of some commercial fertilizers. The ultimate source of virtually all terrestrial phosphorus is from the weathering of minerals and soils in the Earth's crust. Phosphorus is generally available as phosphate, an anion that is not bindable by the cation exchange complex and thus can be easily leached from the soil by rain or runoff. Phosphorus plays the same chemical and biochemical role in plants as it does in all other organisms. It is the main element involved in energy transfer for cellular metabolism and it is a structural component of cell membranes, nucleic acids, and other critical materials. Plants lacking sufficient phosphorus are frequently characterized by phenomena that appear as wound-responses in leaves, such as production of pigmented compounds resulting in darkening or purpling of the leaves. Stunting can also occur, as well as necrotic lesions and other symptoms [2].

2.3. Potassium (K)

Potassium is the primary osmolyte and ion involved in plant cell membrane dynamics, including the regulation of stomata and the maintenance of turgor and osmotic equilibrium. It also plays important roles in the activation and regulation of enzyme activity. Potassium is a soil exchangeable cation and is actively absorbed by plant roots. It is a major component of many soils and is ultimately derived from the weathering of soil parent materials such as potassium-aluminum-silicates in the soil. Potassium though a part of the cation exchange complex, is only weakly held to the soil particles and is highly leachable. Due to plants and other organisms holding potassium as free ions in their cells, once an organism dies, its potassium quickly reenters the soil solution. If other organisms do not quickly take up potassium, it is easily lost from the soil due to leaching and runoff. A loss of potassium is a common result of forest fires, clear-cut harvest methods, and other major disturbances that cause runoff and erosion. Potassium-deficient plants generally form necrotic lesions or more generalized leaf necrosis after a relatively short period of chlorosis. In severely limiting conditions, there can be general bud death. As with nitrogen deficiency, symptoms of potassium deficiency first tend to appear in more mature leaves, as the plant will move potassium to actively growing, younger tissues. Most plants require potassium in fairly high concentration, and as a result, potassium is a common major constituent of commercial fertilizers, particularly in agricultural systems where the removal of plant parts (e.g., fruits) from the site strips potassium from the local cycling system. Sodium, another monovalent cation, can sometimes substitute for potassium in certain plants [2].

2.4. Sulfur (S)

Sulfur is another biologically ubiquitous element, playing critical structural roles in several amino acids and in compounds involved in electron transfers in photosynthesis and respiration. Sulfur is also a structural component of specialized enzymes and related molecules. Sulfur is found in the soil primarily as sulfate and is derived from the weathering of parent soil materials or from byproducts of the human combustion of fossil fuels, which produce the sulfur containing gases hydrogen sulfide and sulfur dioxide. These gases are converted to the sulfuric acid of acid rain. Plants lacking sufficient sulfur often show symptoms such as chlorosis and spindly or stunted growth. Unlike plants deficient in nitrogen or potassium, sulfur-deficient plants generally first show signs of deficiency in the younger, developing tissues because sulfur is not easily translocated within the plant [2].

2.5. Calcium (Ca)

Calcium is a divalent cation that plays important roles in cell wall structure, cell membrane relations, and signal transduction in the plant. Most of these functions are essentially extracellular, occurring in the cell walls rather than within the cell membrane, though calcium's role in cell membrane integrity extends to the intracellular membranes as well. Calcium is derived predominantly from geologic sources from the weathering of soil materials—and is a major ion in the cation exchange complex of the soil. It is fairly uncommon for soils to be deficient in calcium, and most plants seem to grow under conditions with a surfeit of calcium. In plants with insufficient calcium, developing buds, young leaves, and root tips either fail to grow or die, most likely due to cell wall related defects. Calcium is generally made unavailable to plants at low pH (higher acidity), so acidic soils often contribute additional symptoms to the calcium deficiency; many metals become mobile at low pH and are toxic (e.g., aluminum) [2].

2.6. Magnesium (Mg)

Magnesium is another divalent cation but, unlike calcium, its roles are more intimately related to intracellular functions than the predominantly extracellular roles of calcium. Magnesium is the most import mineral in the activation of enzymes. Magnesium is also the central structural element of chlorophyll, and it is involved in the synthesis of nucleic acids. The primary source of magnesium is the weathering of parent materials in soils and, like calcium, it is generally found as a common part of the cation exchange complex or in the soil solution. The solubility of magnesium decreases with increasing acidity and at high pH (alkaline) as well. In the case of low pH, magnesium deficiency will likely occur in conjunction with metal toxicity, due to the increased solubility of metals at low pH. As magnesium plays such a critical role in so many aspects of plant cell biochemistry, there is no single pattern of symptoms for magnesium deficiency. Since magnesium is a necessary component of chlorophyll, plants that have insufficient magnesium often exhibit chlorosis. The symptoms of magnesium deficiency tend to appear first in more mature tissues because magnesium is translocatable within the plant [2].

2.7. Iron (Fe)

Iron is a divalent or trivalent heavy metal, depending on the reduction-oxidation conditions in the soil. It is intimately involved as a structural component of heme-type and other proteins, plays roles in the activation of some enzymes, and is involved in the synthesis of chlorophyll. Iron is found in the soil as various oxides and also in association with various organic molecules. Iron can be limiting in the natural environment due to the unavailability to the plant of the oxide forms of the element. Plants overcome the limitations of iron absorption by both lowering the pH of the soil and increasing the iron solubility and the production of specialized iron-scavenging compounds called siderophores. Siderophores move into the soil, bind with the available iron, and are then reabosorbed by the plant. Once inside the plant, the siderophore is stripped of the iron and then sent back into the soil to secure more iron.

Plants deficient in iron show interveinal chlorosis, first appearing in the younger tissues because iron is not easily translocated within the plant body. In extreme deficiency, even the tissue around the veins becomes chlorotic, and the entire leaf may look pale yellow or white [2].

2.8. Sodium (Na)

Sodium is a micronutrient only for those plants that undergo C4 or CAM photosynthesis rather than C3 photosynthesis. C4 is a specialized form of photosynthesis that is more efficient in hot, dry weather. CAM is a specialized form of photosynthesis that greatly reduces transpirational water loss, typical of cacti and other desert plants. C3 is the most common type of photosynthesis, typical of plants such as maple trees and soybeans. Sodium can also substitute for potassium to a variable degree, depending on the plant species (generally, species that are salt-tolerant can endure a greater rate of substitution). As a monovalent cation, it is a part of cation exchange complex and thus is available in the soil solution. The original source for some sodium is sea salt, but most of the sodium in the soil solution is from salts in the parent soil material. Sodium deficiency is characterized by an inability to photosynthesize properly. In most soils and conditions in the field, a surfeit rather than a dearth of sodium is likely to be the case. Sodium in high concentration in the soil can upset the water potential of the soil solution compared to the roots and thus limit water flow into the plant [2].

2.9. Chlorine (Cl)

Chlorine is necessary for splitting water in photosynthesis, the step that generates oxygen gas breathed by animals. Chlorine is a monovalent anion found largely in soil derived from salts in the parent soil material. It is highly leachable, but is nonetheless available in large amounts, and thus chlorine deficiency is virtually unknown. In the laboratory, it is characterized by the formation of blue-green, shiny leaves that eventually turn a bronze color. In extreme cases, plants wilt or become severely stunted, in addition to having significant chlorosis and necrosis [2].

2.10. Boron (B)

Boron is a neutral micronutrient element, generally present in the soil solution as boric acid. The precise functions of boron in the plant are unknown. It is suggested to have a role in nucleic acid synthesis and general membrane function, as well as in cell wall structural integrity. Plants deficient in boron show general organ brittleness and the apical meristems often die. Roots can also die or become brittle. Such damage often leads to infection by pathogenic organisms, which have little trouble colonizing the already weakened plant [2].

2.11. Manganese (Mn)

Manganese is a heavy metal micronutrient, the functions of which area fairly known. It is involved in the oxygen-evolving step of photosynthesis and membrane function, as well as serving as an important activator of numerous enzymes in the cell, a role it can also share with magnesium in some cases. The symptoms of manganese deficiency largely depend on the species of plant in which the deficiency occurs. In general, manganese-deficient plants form chlorotic and necrotic lesions on the leaves, fruits, or seeds. The distribution of symptoms, whether on younger or older tissues, is dependent on the plant in question [2].

2.12. Zinc (Zn)

Zinc is another heavy metal micronutrient that plays critical roles in many enzymes, often appearing either at the active site of the enzyme or in a position that regulates the enzyme structure. Lack of zinc results in the inability of the plant to make sufficient quantities of these proteins, and thus general growth and extension are limited. Zinc may also be involved in chlorophyll synthesis in some species, and in the synthesis of proteins from DNA. The effects of zinc deficiency are both well known and dramatic. Specifically, plants deficient in zinc often show symptoms known as little leaf and/or rosette growth. In the case of little leaf, the leaves fail to expand to their normal, mature size. Rosette plants are those in which elongation of the stem is almost eliminated, so that all leaves appear to grow from the same place at the base of the stem. Zinc deficiency can also result in stunted growth forms [2].

2.13. Copper (Cu)

Copper is a micronutrient that is heavily involved in electron transfers in energy exchange reactions within the cell, due to its variable oxidation states. It is a component or activator of some enzymes. Copper is a heavy metal found in the soil in association with various other molecules. When found in the plant body, it is typically bound to special molecules within the plant to limit or prevent toxic effects that can arise from high concentrations. Plants deficient in copper often show symptoms of chlorosis or leaf rolling, though there is species-related variability. Woody species sometimes have bark that is blistered, and young shoots may experience dieback [2].

2.14. Molybdenum (Mo)

Molybdenum is a micronutrient specifically for plants that form root nodules with nitrogen-fixing bacteria, though plants that do not form nodules also use trace amounts of it in a protein involved with nitrogen metabolism and uptake. In the case of root nodule–forming species, however, molybdenum plays a structural role in the nitrogen-fixing enzyme nitrogenase. The symptoms of molybdenum deficiency in plants that don't form root nodules include interveinal chlorosis, leaf rolling, and sometimes necrosis. In plants that do form root nodules, molybdenum deficiency results in a loss of productive nitrogen fixation, due to the bacterial need for the element [2].

2.15. Silicon (Si)

Some scientists consider silicon (Si) a micronutrient. Though it not known to be essential, it is accumulated by plants and used in the plant body at a fairly high concentration [2].

2.16. Cobalt (Co)

Cobalt (Co) is an essential micronutrient for plant species that form root nodules [2].

2.17. Nickel (Ni)

Nickel (Ni) is a micronutrient that, while essential, is virtually never limiting or deficient in the natural world. In the rare cases when it is limiting, symptoms include reduction in leaf size, cupping of the leaf, and reduced vegetative growth. It is also a component of a single enzyme, urease. When grown without nickel, plants fail to produce urease in sufficient quantity and can suffer effects of accumulating toxic quantities of urea in the cells [2].

3. Nutrient solutions and its management under soilless hydroponics

Nutrient solutions are critical parts of plant nutrition research. Hydroponics is the science of growing plants in liquid media, rather than in pots of soil [3]. For a hydroponic solution to sustain plant growth, it must provide the required nutrients at appropriate concentrations, and in the correct forms that are available to the plant. Developing a useful hydroponic solution can be a time consuming process. Different plant species may require nutrients in different concentrations, ratios, or chemical forms for efficient absorption [4]. Most plant nutrient solutions, whether used in hydroponics or for watering plants in pots, often employ nutrients at much higher concentrations than they would find in natural soil. The main reason for this approach is to save time in the lab. For example, if there is a high concentration of nutrients present, the solution may need to be changed only once a week instead of once a day, saving considerable time, particularly in an experiment with 500 beakers of plants/ nutrient solution.

Nutrient solutions for growing of some horticultural plants are listed in below.

http://dx.doi.org/10.5772/59478

Total	Cl	SO_4	PO_4	NO_3	meq. L^{-1}
4.6			0.8 0.6	3.2	K
0.2	0.2				Na
5.2				5.2	Ca
1.5		1.5			Mg
0.1				0.1	NH_4
1.9			1.6 0.3		H
13.5	0.2	1.5	3.3	8.5	Total

Table 1. Nutrient solution for growing cucumber [4].

Total	Cl	SO_4	PO_4	NO_3	meq. L^{-1}
5.8			0.8 0.6	4.4	K
0.2	0.2				Na
5.2				5.2	Ca
1.5		1.5			Mg
0.1				0.1	NH_4
1.9			1.6 0.3		H
14.7	0.2	1.5	3.3	9.7	Total

Table 2. Nutrient solution for growing Aloe [1, 3].

Total	Cl	SO_4	PO_4	NO_3	meq. L^{-1}
4.6			0.8 0.6	3.2	K
0.2	0.2				Na
5.2				5.2	Ca
1.5		1.5			Mg
0.1				0.1	NH_4
1.9			1.6 0.3		H
13.5	0.2	1.5	3.3	8.5	Total

Table 3. Nutrient solution for growing garden cress [5, 6]

Total	Cl	SO_4	PO_4	NO_3	meq. L^{-1}
1.8			0.4 0.3	1.1	K
0.1	0.1				Na
2.6				2.6	Ca
0.7		0.7			Mg
0.5				0.5	NH_4
0.95			0.8 0.15		H
6.65	0.1	0.7	1.65	4.2	Total

Table 4. Nutrient solution for growing basil [5].

Total	Cl	SO_4	PO_4	NO_3	meq/l
2.4			0.8 0.6	1.0	K
0.1	0.1				Na
2.0				2.0	Ca
1.5		1.5			Mg
0.5				0.5	NH_4
1.9			1.6 0.3		H
8.4	0.1	1.5	3.3	3.5	Total

Table 5. Nutrient solution for growing anthorium.

4. Preparation of hydroponic nutrient solution for horticultural crops

Hydroponics is a common culture method for production of herbs, vegetables and other crops such as strawberry. It is necessary to determine optimal ranges of elements in nutrient solutions. There are different basic nutrient that was used in soilless culture all around the world. We prefer Quick nutrient solution because this nutrient solution design is simple and easily we are able to changed macronutrient and nutrient solution pH. In below we explain how design a nutrient solution and how prepare stock and nutrient solution.

Step 1.*Research review and total nutrient content selection.*

At the first we have to search about our plant and read all document available. This stage is very important and helps us to design the best nutrient solution. If you don't find any document for your plant then searched for other plant from your plant family that have the highest similarity to your plant. We continue this section with an example (Cucumber). Research review indicated that this plant need high nutrient element with high nitrate to ammonium ratio. So we select 13.6 meq per liter of nutrient solution as total nutrient content (Table 6). In below table each column was related to one anion and each row is related to one cation. So in next step each number was related to one salt that produce from one onion and one cation.

Total	Cl	SO_4	PO_4	NO_3	meq/l
					K
					Na
					Ca
					Mg
					NH_4
					H
13.6					Total

Table 6. Total nutrient content in Quick table.

Step 2. *Total N and NO_3 to NH_4 ratio selection*

After total nutrient selection you have to select total N and NO_3 to NH_4 ratio (Table 7). In some cases it is important to change this level during plant growth especially in different environmental condition like light and temperature. In this step like previous step other researcher published document help you. In cucumber we select total N and NO_3 to NH_4 ratio 8.7 and 86, respectively so our Quick table filled as below.

Total	Cl	SO_4	PO_4	NO_3	meq/l
					K
					Na
					Ca
					Mg
0.1					NH_4
					H
13.6				8.6	Total

Table 7. Ammonium concentration design in nutrient solution

In this step we have to select our nitrogen salt according our availability (Table 8). If potassium nitrate, calcium nitrate and ammonium nitrate were available so our table filled as below:

Total	Cl	SO_4	PO_4	NO_3	meq/l
				3.2	K
					Na
				5.3	Ca
					Mg
0.1				0.1	NH_4
					H
13.6				8.6	Total

Table 8. NO_3 concentration design in nutrient solution.

In fact we select calcium concentration. As you know if concentration of calcium select 2 times more that potassium in nutrient solution they concentration in plants were be equal.

Step 3.*K and pH selection*

After N we must optimize K concentration in our nutrient solution. If we use KH_2PO_4 and K_2HPO_4 we have an opportunity to optimize nutrient pH in our nutrient solution. By increasing KH_2PO_4 concentration the nutrient solution pH decreased. So our table was completed as see in Table 9.

Total	Cl	SO_4	PO_4	NO_3	meq/l
4.6			0.8 0.6	3.2	K
					Na
5.3				5.3	Ca
					Mg
0.1				0.1	NH_4
			1.6 0.3		H
13.6				8.6	Total

Table 9. K and pH design in nutrient solution

In upper table we write 2 numbers for potassium phosphate. If you notice we also have two numbers for H that indicated first number is related to salt with the higher H so the top and below numbers are related to KH_2PO_4 and K_2HPO_4, respectively.

Step 4.*Optimization of other element concentration*

Finally we must select SO_4, Mg, Na and Cl concentration in our table. In below we complete our Table (Table 10).

Total	Cl	SO_4	PO_4	NO_3	meq/l
4.6			0.8 0.6	3.2	K
0.2	0.2				Na
5.3				5.3	Ca
1.5		1.5			Mg
0.1				0.1	NH_4
1.9			1.6 0.3		H
13.6	0.2	1.5	3.3	8.6	Total

Note 1: Micro nutrient have to select separately in another table.

Note 2: Total anion and total cation concentration are equal.

Table 10. Completion of nutrient solution preparation design.

5. Preparation of stock nutrient solution

At first we must calculate our needed salts. In this reason we need molecular weight of each salt. Molecular weight was divided to active capacity of salt. For example active capacity is 2 and 1 for K_2HPO_4 and KH_2PO_4, respectively. The result finally multiplied by table level. In below I calculated each salt concentration.

Nutrients	Conversion	Amount (mg/L)
Macro		
KNO_3	3.2×101	323
$Ca(NO_3)_2$	5.3×82	434.6
NH_4NO_3	0.1×80	8
KH_2PO_4	0.8×136	109
K_2HPO_4	0.6×87	52
$MgSO_4/7H_2O$	1.5×123	184.5
NaCl	0.2×58.5	11.7
Micro		
$(NH_4)_6Mo_7O_{24}/4H_2O$	-	0.05
H_3BO_3	-	1.5
$MnSO_4/4H_2O$	-	2
$CuSO_4/5H_2O$	-	0.25
$ZnSO_4/7H_2O$	-	1
Sequestered Fe 136	-	10

Table 11. Nutrient calculation as mg/L

All macronutrient and micronutrient that listed in Table 11 were prepare separately as stock solution and were used in irrigation solution.

It is better to separate Ca salt and Fe salt from others. So we have 4 stock solutions as below.

1. Stock A = 1/2 KNO_3 + $Ca(NO_3)_2$ + NH_4NO_3

2. Stock B = 1/2 KNO_3 + other macronutrient salts

3. Stock C = Fe salt

4. Stock D = other micronutrient

6. Conclusion

As describe in this chapter nutrient solution is one of the most important step in soilless culture of horticultural crop. By using Quick table we are able to design and change nutrient element in solution. In other hand by using different kind of salt we are able to change pH of nutrient solution. In stock preparation we separate Ca and Fe salts from others kind of salts. It is also possible to separate any salt that not solved completely in stock solution so this is better to test all salts in a small level then prepare stocks solution.

Author details

J.A. Olfati*

Address all correspondence to: jamalaliolfati@gmail.com

University of Guilan, Faculty of Agriculture, Horticultural Department, Rasht, Iran

References

[1] Saliqedar F, Sedaqathour S, Olfati JA. Nutrient solution on Aloin content and other quality characteristics of *Aloe vera*. Journal of Medicinal Plant and Byproducts 2014;1: 59-62.

[2] Wiedenhoeft AC. Plant nutrition. New York. Chelsea house publisher; 2006.

[3] Salighehdar F, Sedaqat-Hoor S, Olfati JA. Effects of four nutrient solutions on vegetative traits of *Aloe vera* L. cv. Austin at six harvest periods. EJGCST 2013;4: 15-27.

[4] Pooyeh F, Peyvast Gh, Olfati JA. Growing media including palm waste in soilless culture of cucumber. International Journal of Vegetable Science 2012;18: 20-28.

[5] Olfati JA, Khasmakhi-Sabet SA, Shabani H. Effect of nutrient solutions on yield and quality of basil and cress using nutrient film technique (NFT). International Journal of Vegetable Science 2012;18: 298-304.

[6] Noori Roudsari O, Peyvast Gh, Alimousavi AR, Ramezani Kharazi P, Olfati JA. Feasibility decrease use of fertilizer on greenhouse cucumber production in soilless culture. Biosciences Biotechnology Research Asia 2008;5(2): 593-600.

Chapter 3

Growing Substrates Alternative to Peat for Ornamental Plants

Giancarlo Fascella

Additional information is available at the end of the chapter

http://dx.doi.org/10.5772/59596

1. Introduction

Nowadays, the search for alternative high-quality and low-cost materials as growing media in horticulture is a necessity due to the increasing demand and rising costs for peat, the most widely used substrates component during the last decades, as well as for its uncertain availability in the near future owed to environmental constraints. The recent and rising interest in waste recycling has resulted in a greater use of organic materials and composts as potting media representing, at the same time, a smart solution for waste disposal problems. In the present chapter, after describing main characteristics and limitations of peats, some alternative organic wastes as substrate components are outlined, comparing their physical and chemical properties to those from peat. Benefits obtained from their use, from an environmental and economic point of view, are briefly discussed. Moreover, three case-studies on peat sustainable substitutes for ornamental plants are reported.

1.1. Peat use in horticulture, characteristics and limits

Among the numerous organic materials used as substrates for soilless cultivation of horticultural crops, peat is currently a major component of containerized mixtures for commercial plant production [1]. Its long-time success is certainly due to the physical properties (slow degradation rate, low bulk density, high porosity, high water holding capacity [WHC]) and the chemical characteristics (relatively high cation exchange capacity, CEC) that makes peat particularly suitable as growing media for a large number of vegetables and ornamentals [2]. Peat is formed as a result of the partial decomposition of plants (Sphagnum, Carex) typical of poorly drained areas (peat bogs), with low nutrients and pH, under low temperatures and anaerobic conditions [3]. Plant species, climatic conditions, harvest and processing methods influence the specific characteristics of peat and its value so different types can be obtained varying on color, texture and degree of decomposition [4]. In particular, some physical properties as water retention and air capacity generally decrease with the increasing of the degree of decomposition. Recently, Prasad and Maher [5] tried to test if determinations of peat colour could be used to predict lignin content and peat stability and they observed a strong correlation between colour parameters and shrinkage as well as with lignin content.

Among different peat typologies, sphagnum moss is maybe the most used for the preparation of soilless substrates because of the light bulk density and the low degree of decomposition [6]. It is obtained from acid bog-plants of the genus Sphagnum and is produced, with high extraction rhythms, in northern regions as Baltic Republics, Finland, Germany and Ireland. Sphagnum peat is usually included in growing mixtures to increase WHC or to decrease the weight of the substrates. It contains 75% fibre at least, consisting of dehydrated remains of leaves and stems of Sphagnum plants; this fibrous structure is characterized by a high surface charge density, with consequent high CEC which helps to reduce leaching of nutrients [7]. Other relevant properties are the high easily available water (EAW) under conditions of container capacity, i.e. after the end of free drainage and the high oxygen diffusion rate. On the other hand, as negative aspect peat can be a conducive substrate for numerous soil-borne diseases and its sterilization does not solve the problem as it leaves a biological vacuum that can be easily filled by pathogenic fungi.

Peat use in horticulture increased during the last decades, resulting in rising costs [8] and generating doubts about availability of this material in the near future due to environmental constraints. In fact, peat mining has been recently questioned because it is harvested from peat lands, highly fragile wetlands ecosystems with a great ecological and archaeological value, included in the list of natural habitats with a potential degradation [9]. Peat also plays an important role in improving groundwater quality, and peat bogs also serve as a special habitat for wild plants and animals. Moreover, these ecosystems represent important carbon dioxide (CO_2) sinks [10]. Thus, the increasing use of peat in horticulture has resulted in a rapid depletion of wetlands, determining the loss of a non-renewable resource and creating a source of greenhouse gases through copious CO_2 release due to the aerobic peat decomposition. For this reason, a global movement has been originated to achieve a sustainable peat use and a smart exploitation of wetlands. Many individual countries (Austria, Switzerland, Germany, Great Britain) have begun to limit the extent of peat mining. Government and commercial peat policies support and encourage the use of sustainable peat substitutes which have to satisfy the specific technical requirements and be readily available in sufficient quantities at reasonable costs.

The increasing demand for soilless substrates for horticultural crop production and the rising environmental concerns about the use of non-renewable resources such as peat as medium has led to the search for alternative materials as constituents of growing mixtures for containerized plants, such as solid organic waste by-products coming from industrial and agricultural activities.

1.2. Growing media alternative to peat

1.2.1. Compost

Compost is a general term describing all organic matter that has undergone a long, thermophilic, aerobic decomposition process calling composting [11]. Composts may vary with raw materials used, and duration and nature of the composting process. The combination of these factors results in a wide range of characteristics (physical, chemical and biological) and qualities of end-product as biological oxygen demand, organic matter and nutrients content, degree of disease suppressiveness [12].

Composts used as growing media are produced from different organic wastes such as sewage sludge, municipal solid waste, animal manure and food-industry waste (sugar cane fibre, olive and grape marc, rice and peanut hulls, cotton gin waste). The latter typology of waste is particularly convenient for composting since it is uniform, rich in organic matter and easily available. Differently from other treatment methods for organic waste (land-filling, incineration) which may cause severe air and/or water pollution (leachates), composting is considered a safer process. It is a method that turns waste in a resource which, if obtained properly, represents a beneficial product for agriculture as able to restore the depleted soil/substrate organic matter [13].

Nowadays composts are widely used as ingredient of growing media for containerized plants for the following reasons: 1) need to find a safe outlet for compost (nonedible plants as ornamentals, forest species) that may be considered not desirable for food crops production; 2) characteristics and performances in containers are similar to peat but with a considerably lower cost; 3) high suppressiveness for many soil borne disease.

Composts used as potted substrates must be stable. Mature compost are more stable than young ones still containing readily biodegradable compounds which can undergo secondary degradation leading to oxygen and N deficiencies in the root zone. As compost stability is not identical to compost maturity, which is a prerequisite for suppressiveness of many root pathogens, mature composts are preferable for growing media preparation.

As regards the physical properties of composts for potted substrates, hydraulic conductivity, as well as air filled porosity (AFP) and EAW should be high. Fast and slow-release of nutrients should be strongly considered as excessive vegetative growth and/or salinity effects may occur, even though high concentrations of phytotoxic ions can be reduced by leaching. It must be reiterated that unless all these requirements are met simultaneously, the compost may fail to serve successfully as a container medium.

Different authors have suggested that some organic materials such as tree bark, sawdust, sludge, and different kind of wastes could be used, after composting, as partial peat substitute [14, 15] as composts may have physical and chemical properties superior or similar to peat because of their higher nutrients availability, not excessive water content, and optimum porosity [16, 17].

The combination of peat and compost in growing media is synergistic: peat often enhances aeration and water retention while compost improves the fertilizing capacity of the substrate.

In addition, organic by-products and composts tend to have porosity and aeration properties comparable to those of peat and, as such are ideal substitutes in propagating media [18]. Because the physical and chemical properties of waste and compost-based media may shift with time and source, these substrates should always be tested for local conditions. Waste-recycling end-products used as composts greatly vary on pH, electrical conductivity (EC), and/or nutrients contents and this variability also depends on the type of collection as well as on the composting process. For this reason, it is important to know the physical, chemical and biological characteristics of each material and to compare them with those required for its use as a growing medium.

On the other hand, though the use of mixtures of compost with peat can minimize the potential poor properties of single materials (heterogeneity, presence of contaminants, immaturity, alkaline pH), the percentage of compost used for potting substrates must be carefully determined to avoid negative effects on plant growth (high soluble salt contents, presence of heavy metals, etc.) [19, 20]. Moreover, disposal of sewage sludge and urban compost may pose an environmental hazard if their heavy metals content is high: in these cases they must be sent to landfills.

1.2.2. Coir dust

Coir dust is produced from the mesocarp tissue, or husk, of the coconut fruit and originates primarily from several tropical countries as Sri Lanka, India, Philippines, Indonesia, Mexico, Costa Rica and Guyana. The Philippines is one of the largest producers of coconuts with >400 million trees, Sri Lanka annually produces from 350,000 to 500,000 tons of new husk [21]. With this level of production, large volumes of coir dust are potentially available to horticultural markets.

The husk contains approximately from 60 to 70% pith tissue with the remainder being fibre of varying lengths. Husks may be soaked in water to soften them and facilitate grinding. After grinding of the husk, the long fibers are removed and used for various industrial purposes such as rope and mat making. The remaining material, composed of short and medium length fibers as well as pith tissue, is commonly referred to as waste-grade coir. The waste-grade coir may be screened to remove part of the fiber, and the remaining product is referred to as coir dust which is more stable while fibers tend to undergo secondary decomposition in the growth medium [22]. During composting, hemicellulose, cellulose and partially lignine components are decomposed, causing an increase of C/N ratio, CEC and humic acid content, as well as of some physical properties like total porosity, EAW and water buffer capacity, but a decrease of AFP. After composting, coir dust is allowed to dry to a specific moisture level and is then compressed into bales, wrapped, and shipped. The source, the moisture level and the compression pressures often differ among producers so coir is not a uniform material resulting in a large variability of end-product. With the addition of water, coir dust expands to 5 to 9 times its compressed volume [23].

Coconut coir dust (CD) is widely used, alone or mixed with other materials, as an alternative growing medium for soilless cultivation of vegetables, cut flowers and potted plants as it evidenced growth performances similar to that of peat. Coir can also be used as rooting medium for cuttings under mist because of the presence of root-promoting substances. Evans et al. [24] examined the chemical and physical properties of CD from numerous sources and reported that properties were generally within acceptable ranges except for EC and chloride, which often exceeded recommended levels. Coir physical properties usually varied in according to the quantity of fibrous particles included, so increasing fibre is generally associated with increased porosity and decreased bulk density and WHC.

Coir dust characteristics were also investigated by other authors who reported this material of plant origin as suitable for use in substrates and an effective substitute for sphagnum peat moss for many container crops [25, 26, 23]. In fact, it may present some chemical and hydrological features (organic matter content, CEC, water retention) similar to peat, but with a higher pH and durability. Shrinkage was found to be lower compared to sphagnum moss and higher than in Irish peat moss.

Nevertheless, literature on main physical and chemical characteristics of coir dust is sometimes contradictory: discordances among references can be linked to the heterogeneity of the material which presents different features related to the source and fibre size. pH (in water) ranged between 5 and 7, so higher than peat and suitable for neutrophil crops, without need to use adequate adjustments ($CaCO_3$). CEC ranged from 30 and 100 meq/100g, values similar to that of brown peat, so with a high buffer capacity. Sometimes a high salinity occur due to a high content of K, Na and Cl as coconut palms live near seashores. EC measured on fresh coir fibre ranged between 0.3 and 2.9 dS m^{-1}, according Sonneveld method (1:1.5 v/v), whereas an EC lower than 0.5 dS m^{-1} is optimal for a substrate component. Soluble salts level affect the quality of coir dust: high salinity and, in particular, excessive content of Na and Cl may cause severe problems according to plant species and growth stage. Evans and Stamps [25] reported that coir dust with a Cl content of 600-700 mg L^{-1} may provide high-yield results if a leaching was applied to plants.

Air content at pF1 is similar to that of blonde peat [27], but extremely different values (from 9 to 92% of total volume) have been recorded from other authors. Water retention capacity seems to be higher than sphagnum peat: according Evans et al. [24], coir dust retention is about 750-900% of its weight, while that of peat is about 400-800%. Contrasting information are however present in literature: Prasad [28] refers about a higher water retention in peat than in coir. Changes of physical characteristics of coir dust are slower than those of sphagnum peat, indicating a higher bio-stability during use (cultivation).

1.2.3. Biochar

Biochar (biologically derived charcoal) is a fine-grained and porous substance produced by pyrolysis, a 300-500°C thermo-chemical process where waste biomass is heated in the absence of oxygen [29]. As results, bio-oil, synthesis gas and black carbon (biochar) are obtained. It can be obtained from different feedstocks (tree wood, grape wine marc, olive cake, chicken manure). Also known as Amazonian Dark Earth or Terra Preta de Indio, biochar is a stable solid material originally obtained from the carbonization of biomass which endured in soil for hundreds of years. It is characterized by the presence of low-temperature charcoal in high concentrations, high quantities of organic matter (plant residues, manures, bones), and nutrients. It also shows high levels of microorganism activities.

Soil application of biochar can be used to overcome some of the limitations faced during land farming, thereby providing a supplementary management option in addition to other organic materials and having many environmental and sustainability advantages over manures and composts. In fact, it is a porous material with a high inner surface area which helps to retain more water and increase saturated hydraulic conductivity of top soils [30]. Biochar may improve the physical structure of the soil and can also modify soil hydraulic properties: as its pore size is relatively fixed, biochar increases available moisture in sandy soils while has a neutral effect in medium textured ones and decreases moisture availability in clay soils. Glaser et al. [31] observed that biochar-enriched Terra Preta had a WHC 18% higher than the adjacent soils. Biochar seems able to decrease nutrient leaching thus enhancing nutrient availability. Moreover, its CEC is consistently higher than that of the whole soil: the concentration of negative charges on biochar surfaces increases with age as well as the adsorption of charged organic matter. Field experiments on biochar application in different soils and crops have been conducted, and describing positive yield responses [32, 29] and attributing them to the effects of biochar on nutrients availability (i.e. nutrient savings in terms of improved fertilizer use efficiency). Therefore, biochar can enhance soil fertility, increase agricultural productivity and provide protection against some foliar and soil-borne diseases.

Recently, Lehmann et al. [32]and Steiner et al. [33] introduced the concept of converting residues to biochar as an alternative agricultural method to reduce CO_2 emissions. In fact, soil application of biochar may have the greatest potential for the long-term sequestration of carbon (C) as it can remain in the soils for many hundreds of years, due to its stable structure and complex aromatic polycyclic form, thus enhancing the resistance of C to microbial decay and replenishing the scarce carbon stocks. For these reasons, incorporating biochar into soil is currently considered as an interesting option to reach mitigation targets like agricultural management able to reduce greenhouse gases (atmospheric CO_2 concentrations) [34, 35]. Increased soil C sequestration, through biochar addition, can improve soil quality because of the vital role that this element plays in chemical, biological, and physical processes.

Aside from the lack of commercial biochar available to farmers and legislative barriers that prevent it being applied to land (e.g. in Europe) due to the main uncertainty about its long-term performance, widespread adoption of biochar application from a large variety of feedstocks is partially hampered by the unpredictability of plant growth response across different systems. As with many agricultural practices, biochar is reported to result in positive, negative and neutral effects on productivity. Direct comparison of plant growth outcomes is often difficult due to the high variation in numerous experimental parameters including the particular biochar used (feedstock and pyrolysis conditions), the studied plant system (annual/perennial, vegetable, ornamental, etc.) and the growth resources provided (soil type, nutrient availability, moisture, etc).

Until now, numerous studies on biochar agricultural use have been conducted on its application on soil but few researches on the utilization as growing medium for potted plants have been carried out [36, 37, 38, 39]. The positive characteristics of biochar as soil ameliorant (enhancing CEC, reducing nutrient run-off, improving water retention capacity, providing suitable conditions for micro-organisms) could be exploited for using it as a substrate component, together or as alternative to peat, for containerized plants.

2. Case-studies on peat substitutes for ornamental plants

2.1. Sphagnum peat and coir dust as growing substrates for *Euphorbia × lomi* hybrids in soilless culture

2.1.1. *Aim of the study*

In order to evaluate the performances of sphagnum peat and coir dust as growing media for ornamentals, a study of soilless cultivation of *Euphorbia × lomi* Rauh (an interspecific hybrid recently introduced to the Mediterranean countries as a new floral crop) using two organic substrates was carried out, collecting data on growth and production and considering possible technical problems for plant management. In fact, the possibility to grow Spurge family plants in soilless culture with organic substrates could be interesting to maintain mother-plants of these genotypes in optimal health conditions during a mass propagation process, evaluating their vegetative and productive behavior. In fact, it is well-known that one of the numerous advantages of this innovative technique is to limit problems associated with the soil as soil-borne diseases.

2.1.2. *Materials and methods*

The study was carried out in a double-span polyethylene -covered 540 m^2 greenhouse (28°C day/14°C night). Mother plants of the Thai cultivars 'Nam Chok' and 'Sabckaeron Suk' were grown in polypropylene benches (720 L) filled with two growing media composed of sphagnum peat/perlite (1:1, v/v) and coconut coir dust/perlite (1:1, v/v) in an open-loop system with no recirculating solution. The physical and chemical characteristics of the organic substrates were determined according De Boodt et al. [40] and Sonneveld et al. [41], respectively.

A split-plot experimental design with two substrates as the main plot and two cultivars as subplots with three replications and 20 plants per replication was used. Plants were transplanted in double rows (row spacing of 0.4 m) with a final density of 6.2 plants m^{-2}. Water and nutrients were supplied by a drip system controlled by a computer. Irrigation scheduling was performed using electronic low-tension tensiometers that control irrigation on the basis of substrate matric potential [42]. Plants were daily fertigated at 2 L h^{-1} one to five times during the growing cycle. The duration of each fertigation was adjusted when the drainage exceeded the range of 10 to 20%. The composition (mg L^{-1}) of the supplied nutrient solution was as follows: 150 N total (NO_3+NH_4), 50 P, 200 K, 120 Ca, 30 Mg, 1.2 Fe, 0.2 Cu, 0.2 Zn, 0.3 Mn, 0.2 B, and 0.03 Mo. The pH and the EC were maintained at 6.0 and 2.0 dS m^{-1}, respectively.

Plant height, stem diameter, total number (basal and lateral) of shoots per plant, and number of cuttings suitable for rooting (with average length of 8-12 cm) harvested per plant, were recorded for a 12-month period. Water absorption was calculated from the difference between the volume of nutrient solution applied and the volume of collected drainage. Nutrients content in the root zone (uptaken by roots and retained by substrate) was determined by photometric test as the difference between the concentration of each element in the given solution and in the collected drainage.

Collected data were subjected to two-way analysis of variance and means were separated according to Duncan's multiple range test at $p \leq 0.05$.

2.1.3. Results and discussion

As regards physical and chemical characteristics of the two organic substrates, similar values were recorded on bulk density and total porosity, whereas air content was higher in coir dust/perlite than in sphagnum peat/perlite (48.1 and 34.5%, respectively) (Table 1). Peat-based substrate showed higher WHC (58.6 and 47.2%, respectively) and EAW (20.1 and 13.4%) than those measured in coir dust, which was also characterized by a higher pH. EC was similar in both media, while CEC was higher in peat/perlite than in the coir dust-based substrate (55.2 and 36.1 meq/100 g, respectively) (Table 1).

Plants grown in sphagnum peat/perlite showed a similar height than those cultured in coir dust/perlite (51.7 and 48.2 cm, respectively) as well as a similar basal stem diameter (Table 2). No significant differences between substrates were recorded as regards shoots production: an average amount of 18.3 shoots plant^{-1} was obtained regardless of the growing medium (Table 2). A higher number (10.4) of cuttings suitable for rooting was produced from plants cultivated in peat-based substrate compared with that (5.6) from hybrids grown in coir dust (Table 2). Higher water absorption was recorded from plants grown in peat/perlite (265.2 mL plant^{-1} day^{-1}) than those cultivated in coir dust mixture (153.4 mL plant^{-1} day^{-1}) (Table 3). Plants in sphagnum peat/perlite evidenced higher macro- and micronutrients content in the root zone compared to that recorded in coir dust (Figure 1).

The influence of the two organic mixtures on plant growth, water and nutrients absorption are most likely correlated to their physical and chemical properties, which were previously described by other authors [24, 43, 44] who referred that the sphagnum peat and coir dust, though showing some similarities, significantly differ on important chemical and hydrological characteristics: coir dust evidenced higher porosity and air content and lower total and EAW capacity than peat [45].

In our case-study, the similar growth performances (absence of differences recorded on plant height, stem diameter, and shoot total production) recorded in plants cultivated in peat/perlite and coir dust/perlite, seem to suggest that *Euphorbia* × *lomi* hybrids can be grown in both substrates, corroborating the thesis according with coir dust is considered one of the most important peat substitute as organic medium for soilless cultivation of ornamental plants.

Substrate characteristics	Sphagnum peat/Perlite	Coir dust/Perlite
Bulk density (g cm^{-3})	0.12 a^z	0.11 a
Total porosity (% v/v)	93.1 a	95.3 a
Air content (% v/v)	34.5 b	48.1 a
Water holding capacity (% v/v)	58.6 a	47.2 b
Easy available water (% v/v)	20.1 a	13.4 b
pH (on water extract 1:1.5 v/v)	5.3 b	6.0 a
Electrical conductivity (dS m^{-1})	0.5 a	0.6 a
Cation exchange capacity (meq/100g)	55.2 a	36.1 b

^{z}In any row, means followed by different letters are significant at p≤0.05 (DMRT)

Table 1. Physical and chemical characteristics of the sphagnum peat and coir dust-based growing media of *Euphorbia* × *lomi* soilless plants

Substrate mixture	Plant height (cm)	Stem diameter (cm)	Shoots (n. plant^{-1})	Cuttings (n. plant^{-1})	Water absorption (mL plant^{-1} d^{-1})
Sphagnum peat/Perlite	51.7 a^z	3.2 a	21.0 a	10.4 a	265.2 a
Coir dust/Perlite	48.2 a	3.0 a	15.5 a	5.6 b	153.4 b
Significance	ns	ns	ns	*	**

^{z}Within a column, means followed by different letters are significant at p≤0.05 (DMRT)

$^{ns, *, **}$ = nonsignificant or significant at $p \leq 0.05$ and ≤ 0.01, respectively.

Table 2. Effects of organic soilless substrates on *Euphorbia* × *lomi* plant height, stem diameter, shoots and cuttings production, and water absorption

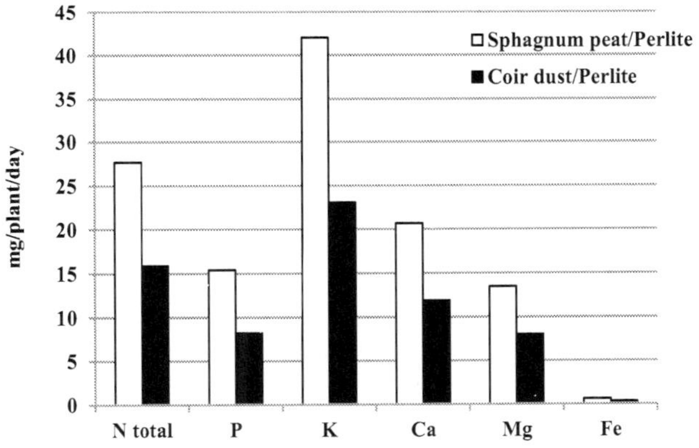

Figure 1. Nutrients content (mg plant^{-1} day^{-1}) in the root zone of *Euphorbia* × *lomi* plants as affected by organic soilless substrates

2.2. Changes in physico-chemical characteristics and growth performances of a coir dust-based substrate during a long-term cultivation of cut rose plants

2.2.1. Aim of the study

Coconut coir dust is frequently used as organic medium, singularly or mixed with inert materials (perlite), in soilless systems for vegetable crops but is rarely employed for pluriannal culture of ornamental species. The possibility of finding affordable growing substrates suitable for long-term cultivation of cut flowers could allow for a reduction of prime costs for growers and avoid a short turn-over of plants and substrates. Few reports on the reuse of growing materials (pluriannal cycles on the same substrates) for ornamental species are available in literature and less on a prolonged period of culture of hydroponically grown roses.

Most physical characteristics described for coir dust have been recorded at the beginning of a crop or shortly thereafter, but is highly probable that these properties would change over time as coir resulted in NO_3 depletion during plants cultivation due to microbial decomposition. Therefore, it is important to determine the physical characteristics of a substrate over a crop period rather than just prior to production.

The aim of this study was to test the changes in the physical and chemical properties of a coir-based growing medium during a three year-soilless cultivation cycle of cut roses, collecting data regarding the evolution of substrate characteristics as well as rose yield and quality response.

2.2.2. Materials and methods

The study was conducted in an unheated (28 °C day/14 °C night) single-span EW oriented greenhouse (25 ×8 m) with steel structure and polyethylene cover (thickness 0.15 mm). Plants of *R. hybrida* cultivars 'Dallas' and 'Red France', grafted on *R. indica major* rootstock, were grown in 80 L polyethylene bags filled with a mixture of coconut coir dust and perlite (3-5 mm diameter) (1:1, v/v) in a semi-closed hydroponic system. Each bag (100 × 50 cm) supported 10 plants of the same cultivar with a final density of 4.5 plants m^{-2}.

A completely randomized blocks experimental design was used; each treatment (the two cultivars) was replicated 3 times; each replicate was a group of 30 plants (3 bags) leading to a total of 180 plants (30 plants × 3 replications × 2 cultivars). All the plants were cultivated following the 'arching' technique according which weaker and unmarketable stems were bent horizontally in order to promote basal shoot formation and to increase plant canopy and light interception [46, 47].

Water, macro and micronutrients were supplied to plants via a drip-system (1 dripper $plant^{-1}$, 2 L h^{-1}) which was automatically controlled by a fertigation computer. The nutrient solution had the following composition (mg L^{-1}): 180 N total (NO_3+NH_4), 50 P, 200 K, 120 Ca, 30 Mg, 1.3 Fe, 0.2 Cu, 0.2 Zn, 0.3 Mn, 0.2 B and 0.03 Mo. The pH and the EC were maintained at 5.8 and 1.8 dS m^{-1}, respectively.

Irrigation scheduling was performed using electronic low-tension tensiometers that control irrigation on the basis of substrate matric potential. The number of daily irrigations varied from 3 to 6 (corresponding to 0.4 and 1.5 L $plant^{-1}$ day^{-1}, respectively). The duration of each delivery was adjusted when the leachate fraction exceeded, for each growing material, the range of 15-25%. This fraction was calculated by collecting the drainage solutions.

The main physical properties (bulk density, total pore space, air content, WHC and EAW) and the chemical characteristics (pH, EC and CEC) of the coir dust-based substrate were determined according to De Boodt et al. [40] and Sonneveld et al. [41], respectively, at the beginning and at the end of the trial. Four bags were randomly selected and analyzed before planting and another four were selected and analyzed after 36 months and removal of the 40 plants.

Nutrient content in the root zone was determined by a photometric test and calculated, at the end of the first year of cultivation and at the end of the third one, as the difference between the concentration of each element in the supplied solution and in the collected leachate.

Rose stems were harvested by cutting to the second 5-leaflet leaf from their origin. Parameters as number of stems plant^{-1}, stem length, basal stem thickness and flower bud height and width were recorded throughout the trial.

Data collected over the 36 month-period were subjected to one-way analysis of variance and means were separated at $p \leq 0.05$ using Duncan's multiple range test.

2.2.3. Results and discussion

Numerous changes in physical and chemical properties of the coir dust-based substrate were recorded during the 36 month-growing period: bulk density significantly increased after 3 years of cultivation, whereas total pore space (TPS) moderately decreased (-6.2%) and air content significantly decreased (-18.3%) (Table 3). In the same period, WHC of the organic mixture increased (+15.6%) and EAW moderately improved (+6.2%) (Table 3). During the growing period, the pH of the substrate did not vary considerably, whereas the EC significantly increased (Table 3); no difference in the CEC was evidenced from the beginning to the end of the experiment. A general decrease in the content of macro and micronutrients in the root zone of the growing medium was also shown from the 1st to the 3rd year of rose cultivation (Table 4).

With regard to the influence of the length of the growing period on flower yield, prolonged cultivation was characterized by an increase in yield (+61%) during the 2nd year and by a decrease (-29%) in the 3rd one (Figure 2). Rose plants averagely produced 15.5 cut stems during the 1st year of culture, 25.3 in the 2nd one and 18.0 in the 3rd, respectively. Significant differences were also observed between cultivars as 'Dallas' evidenced a higher flower production than 'Red France' (21.5 and 17.8 stems plant^{-1}, respectively) (Figure 2). Triennial rose yield response of our case-study agrees with the outcomes recorded in a 2.5 year-trial with gerbera cultured on different growing media [48].

As regards the annual variations of quality traits of cut flowers, stem length showed constant values (average 65.4 cm) during the first two years of cultivation, but slightly decreased in the third one (60.0 cm) (Table 5). A progressive decrease of stem thickness was observed from the beginning (8.6 mm) to the end (6.8 mm) of the experiment. Flower bud height increased from the 1st to the 2nd year (from 5.1 to 5.6 cm) of cultivation but reduced in the 3rd one (4.8 cm) (Table 5). A progressive increase of bud width (from 4.6 to 6.1 cm) was yearly recorded all over the study.

Different yields and quality performances of soilless roses grown in coir dust-based medium during the three-year case-study are most likely linked to the physical and chemical properties of coir dust and to their evolution throughout the cultivation period. Actually, numerous changes in main physical and hydrological characteristics of the tested mixture occurred during the 36-month culture: bulk density increased whereas TPS and air content decreased, WHC and EAW increased. These outcomes agree with those reported by Nowak and Strojny [49] during a 1.5 year-cultivation of gerbera in different growing media.

As conclusive remarks, this case-study indicates that coir dust is highly suitable as organic growing medium for cut rose production during a three-year soilless culture in a south Mediterranean region. This material of plant origin, mixed with perlite, resulted in high yield and quality with an adequate physical and chemical stability over time (high WHC, CEC and nutrients content in the root zone, essential factors for successful plants performances in the extreme [summer] greenhouse conditions), sufficient to ensure a relatively long turn-over of crops and substrates.

Substrate characteristics	1st year	3rd year
Bulk density (g cm^{-3})	0.13 b[z]	0.24 a
Total pore space (% vol.)	95.2 a	89.0 b
Air content (% vol.)	58.5 a	40.2 b
Water holding capacity (% v/v)	33.2 b	48.8 a
Easy available water (% v/v)	11.2 b	17.4 a
pH	6.4 a	5.3 b
Electrical conductivity (dS m^{-1})	0.6 b	2.2 a
Cation exchange capacity (meq/100 g)	45.2 a	36.1 a

[z]In any row, means followed by different letters are significant at $p \leq 0.05$ (DMRT)

Table 3. Physical and chemical characteristics of coir dust/perlite recorded at the beginning and at the end of the three years of soilless rose culture.

Nutrients	1st year	3rd year
N	118.4 a[z]	90.5 b
P	58.0 a	41.6 b
K	130.2 a	107.1 b
Ca	72.9 a	64.0 a
Mg	33.2 a	22.3 b
Fe	1.2 a	0.6 b

[z]In any row, means followed by different letters are significant at $p \leq 0.05$ (DMRT)

Table 4. Nutrient content (mg L^{-1}) in the root zone recorded at the end of the 1st and of the 3rd year of cultivation in the coir dust-based substrate.

Growing years	Stem length (cm)	Stem thickness (mm)	Bud height (cm)	Bud width (cm)
1st year	65.2 a[z]	8.6 a	5.1 ab	4.6 b
2nd year	65.6 a	7.3 ab	5.6 a	5.8 a
3rd year	60.0 a	6.8 b	4.8 b	6.1 a
Significance	ns	*	*	*

[z]Within a column, means followed by different letters are significant at p≤0.05 (DMRT). ns, *, = not significant, significant at P≤0.05.

Table 5. Annual variations of cut roses qualitative characteristics during the three-year growing cycle in coir dust-based substrate.

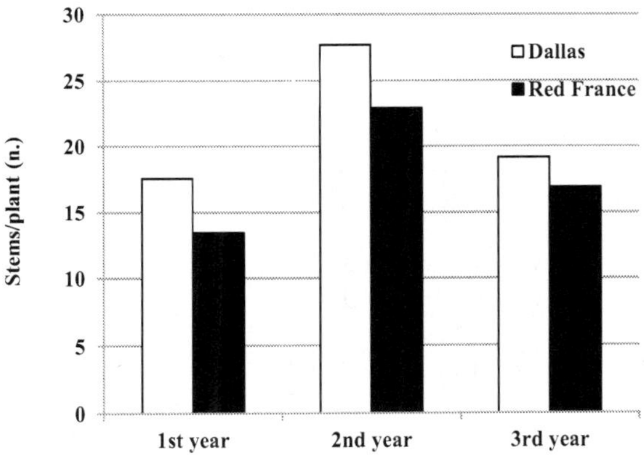

Figure 2. Annual variations of cut roses production of two cultivars during the three year growing cycle in coir dust-based substrate.

2.3. Conifers wood biochar as peat reduced-growing substrate for containerized ornamental plants

2.3.1. *Aim of the study*

The present study deals with the use of biochar made from conifers wood as a growing medium for containerized *Euphorbia* × *lomi* in order to reduce peat use in horticulture. The scopes of this work were to evaluate the main physical and chemical properties of potting substrates composted with decreasing content of sphagnum peat and increasing percentages of biochar, and to observe the influence of these materials on the growth and ornamental characteristics of flowering potted plants.

2.3.2. *Materials and methods*

The study was conducted in an unheated single-span EW oriented greenhouse (25 × 8 m) with steel structure and polyethylene cover (thickness 0.15 mm). Plants of *Euphorbia* × *lomi* Rauh cv. 'Serena' were grown in plastic pots of 13 cm diameter (vol. 1 L) filled with different mixtures (v/v) of sphagnum peat and conifers wood biochar (100% peat – 0% biochar, 85% peat - 15% biochar, 70% peat - 30% biochar, 55% peat - 45% biochar, 40% peat - 60% biochar, respectively). Used biochar derived from pyrolysed (at 450 °C for 48h) trunks and branches of silver fir, larch, spruce, black pine, and Scots pine trees.

Water, macro and micronutrients were supplied to plants through a drip fertigation system (1 dripper plant^{-1}, 2 L h^{-1}) controlled by a computer. All plants were fed with the same nutrient solution which had the following composition (mg L^{-1}): 180 N total (NO_3+NH_4), 50 P, 200 K, 120 Ca, 30 Mg, 1.2 Fe, 0.2 Cu, 0.2 Zn, 0.3 Mn, 0.2 B. The pH and the EC of the nutrient solution were maintained at 6.0 and 2.0 dS m^{-1}, respectively.

Main chemical (pH and EC) and physical characteristics (bulk density, TPS, air and water content) of the tested substrates were analyzed according Sonneveld et al. [40] and De Boodt et al. [41], respectively. Plant growth (plant height, stem diameter, leaf area, root length, dry biomass and its allocation) and ornamental traits (number of leaves, flowers, and shoots, number of marketable plants) were monitored during the trial. Dry weight of the biomass was determined after 72h in a 100°C air-forced oven when harvested tissues reached a constant value. Leaf area (LA) was measured using a digital area meter. Leaf chlorophyll content of three randomly selected leaves of all plants in each experimental unit was measured with a chlorophyll meter and expressed as SPAD unit. Percentage of marketable plants was determined as the amount of potted plants with a high ornamental value (compact habit, presence of three open inflorescences at least, absence of leaf chlorosis, etc.) at the end of the trial (3 month-cultivation).

A completely randomized blocks design with 3 replications per treatment was used; each replication consisted of 20 plants. Collected data were subjected to one-way analysis of variance and means were compared using Duncan's Multiple Range Test at 5% of probability by using a statistical software package.

2.3.3. *Results and discussion*

Addition of conifers wood biochar significantly affected chemical characteristics of the growing substrates as pH increased (from 5.7 to 7.9) with the increase of biochar content, while higher value of EC was recorded in the substrate with 100% peat (Table 6). Biochar addition also influenced physical characteristics of the growing media as bulk density increased together with the increase of biochar content (from 310 to 525 g L^{-1}), while TPS moderately increased (from 77.5 to 90.4%) (Table 6). Air content did not significantly varied among treatments whereas water content moderately decreased (from 58.7 to 48.3%) as biochar content in the substrates increased. Vaughn et al. [50], during an experiment with wheat straw and wood biochar for peat moss replacement in soilless substrates, referred that both biochars (at rates of 5, 10, and 15%, v/v) had significantly higher pH, EC and bulk density than peat

http://dx.doi.org/10.5772/59596

moss. Our results partially differed with those obtained from Dumroese et al. [37] who reported that pelletized wood-derived biochar used in soilless substrate performed well when substituted for peat at a rate of 25% (v/v) only, but at higher levels (50, 75 and 100% pellets) proved unsatisfactory, possibly due to high C/N ratios and bulk densities, and swelling of the substrates after the addition of water.

As regards *Euphorbia* × *lomi* growth, biochar content did not affect plant height, leaves and shoots production averaging 16.6 cm, 90.9 leaves plant^{-1} and 13.1 shoots plant^{-1}, respectively, across all treatments (Table 7). Stem diameter was higher (18.5 mm) in plants grown with 60% biochar as well as for leaf area (1505.0 cm^2). No significant differences among substrates were recorded on leaf chlorophyll content (SPAD values). Flower production and root length were influenced by biochar content of the growing media as higher values (2.6 inflorescences plant^{-1} and 18.1 cm, respectively) were observed in plants grown with 45% and 60% biochar (Table 7).

Biochar content of the growing substrates significantly affected biomass production and its allocation as higher dry weight of plants (26.0 g) were recorded in Euphorbia grown with lower peat percentage (Figure 3). Biochar also influenced the number of marketable potted plants obtained at the end of the trial as an increase was observed (from 24.3% to 56.7%) by increasing biochar content in the growing media (Figure 4). Results from our case-study are more encouraging than those recorded by Vaughn et al. [50] who reported that straw and wood biochar addition to peat in potted tomatoes and marigolds significantly increased plant heights in all treatments but had only a minor or even no effect on dry weights.

Biochar content[z]	pH	EC (dS m^{-1})	Bulk density (g L^{-1})	Total pore space (% v/v)	Air content (% v/v)	Water content (% v/v)
0%	5.7 b[y]	46 a	310 e	77.5 b	32.3 a	58.7 a
15%	6.4 ab	16 b	350 d	80.1 b	29.2 a	57.6 a
30%	6.7 ab	15 b	420 c	82.2 ab	27.3 a	53.4 a
45%	7.3 a	24 b	485 b	85.7 ab	34.0 a	46.1 b
60%	7.9 a	25 b	525 a	90.4 a	32.1 a	43.3 b

[z]Substrate mixture contain 100% peat – 0% biochar, 85% peat - 15% biochar, 70% peat - 30% biochar, 55% peat - 45% biochar, and 40% peat - 60% biochar.

[y]Within a column, means followed by different letters are significant at p≤0.05 (DMRT)

Table 6. Effect of biochar content in the growing substrates on main chemical and physical characteristics

As final remarks, results of this study seem to indicate a high suitability of conifers wood biochar as an alternative to peat for growing media component of *Euphorbia* × *lomi* containerized plants. In fact, using a substrate composed with 60% biochar and 40% sphagnum peat is possible to obtain marketable plants with high ornamental value after a 3 month-cultivation period. Obviously, other researches are needed in order to evaluate and/or to confirm the performances of biochar obtained from other biomass feedstocks and with other ornamental species.

Biochar content[z]	Plant height (cm)	Stem diameter (cm)	Leaves (n. plant^{-1})	Leaf area (cm^2)	SPAD	Flowers (n. plant^{-1})	Shoots (n. plant^{-1})	Root length (cm)
0%	16.4 a[y]	12.3 b	92.3 a	1114.0 b	49.5 a	1.2 b	13.8 a	12.7 b
15%	15.8 a	13.5 b	97.8 a	1035.0 b	42.6 b	1.2 b	14.7 a	13.8 b
30%	17.5 a	13.8 b	93.0 a	1245.2 ab	45.3 ab	1.8 b	11.7 a	17.5 a
45%	15.9 a	13.8 b	85.5 a	1377.0 ab	44.1 ab	2.7 a	12.2 a	18.3 a
60%	17.3 a	18.5 a	86.0 a	1505.0 a	46.8 a	2.5 a	13.2 a	18.0 a

[z]Substrate mixture contain 100% peat – 0% biochar, 85% peat - 15% biochar, 70% peat - 30% biochar, 55% peat - 45% biochar, and 40% peat - 60% biochar.

[y]In any column, means followed by different letters are significant at p≤0.05 (DMRT)

Table 7. Growth and ornamental characteristics of *Euphorbia × lomi* containerized plants as affected by biochar content in the growing substrates

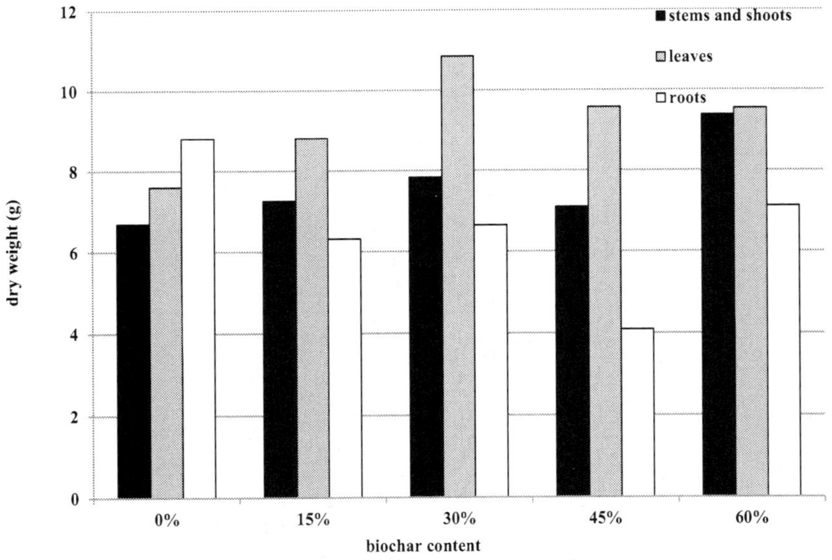

Figure 3. Influence of biochar content in the growing substrates on dry matter allocation of *Euphorbia × lomi* containerized plants. Substrate mixture contain 100% peat – 0% biochar, 85% peat - 15% biochar, 70% peat - 30% biochar, 55% peat - 45% biochar, and 40% peat - 60% biochar.

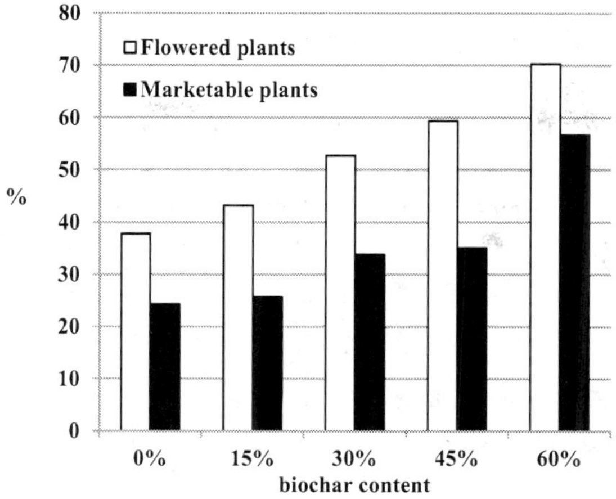

Figure 4. Influence of biochar content in the growing substrates on flowered and marketable potted plants (%) of *Euphorbia × lomi* after 3 months of cultivation. Substrate mixture contain 100% peat – 0% biochar, 85% peat - 15% biochar, 70% peat - 30% biochar, 55% peat - 45% biochar, and 40% peat - 60% biochar.

3. Conclusion

Results reported in the numerous studies previously conducted on peat substitutes, as well as outcomes from the three case-studies above described, confirm that many organic materials, after proper composting, may be used as soilless substrates components for ornamental crops. Some by-products obtained from waste recycling of human activities, agricultural and food industry, and/or energy production processes represent valid alternative to peat, partially or totally, as constituents of growing media for cut flowers and flowering potted plants because having adequate physical and chemical properties and high contents of nutrients. However, their use as substrates depends on the species to be cultivated, as the EC and potentially toxic element accumulation are the main limiting factors. Therefore, the percentage of these waste components in the final substrate is extremely important, with the aim to minimize potential hazards, especially salinity. The evaluation of the beneficial (root zone improvement, nutrients input) and non-beneficial effects (salinity, heavy metals) of organic residues–peat mixtures on growth and yield of ornamentals have to be considered, in order to optimize their wide application. Balanced proportions of many of these materials combined with other compounds (inert or organic), instead of using singularly, could allow to avoid possible negative effects on plant growth and production. As described before and as reported by many authors, coconut coir dust provided higher performances on ornamental plants when combined with inert materials like perlite at 40-60% ratios of substrate final volume depending on plant species, irrigation and nutrient managements; conifers wood biochar may be used as growing medium even with no previous composting and showed best yield and quality results when mixed with specific amounts of sphagnum peat.

Author details

Giancarlo Fascella*

Address all correspondence to: fascella@excite.it

Consiglio per la Ricerca e la Sperimentazione in Agricoltura, Unità di Ricerca per il recupero e la valorizzazione delle Specie Floricole Mediterranee, Bagheria, Palermo, Italy

References

[1] Sprinkles C., Bachman G.R. 1999. Germination of woody plants using coir as a peat alternative. SNA Research Conference, 44

[2] Bohlin C., Holmberg P. 2004. Peat dominating growing medium in Swedish horticulture. Acta Horticulturae 644, 177–181.

[3] Raviv M., Wallach R., Silber A., Bar-Tal A. Substrates and their analysis. In: Savvas D., Passam H. (eds.) Hydroponic production of vegetables and ornamentals. Athens: Embryo Publications; 2002. p25-101.

[4] Von Post L. The geographical survey of Irish bogs. The Irish Naturalists' Journal (1937); 210-227.

[5] Prasad M., Maher M.J. 2013. Colour of peat as an indicator of chemical, biological and physical properties of peats. Acta Horticulturae 1013, 89-94.

[6] Heiskanen J. Variation in water retention characteristics of peat growth media used in tree nurseries. Silva Fennica (1993);27 77-97.

[7] Li Q., Chen J., Caldwell R.D., Deng M. Cowpeat as a substitute for peat in container substrates for foliage plant propagation. HortTechnology (2009);19(2) 340-345.

[8] Abad M., Noguera P., Bures S. National inventory of organic wastes for use as growing media for ornamental potted plant production: case study in Spain. Bioresource Technology (2001);77 197–200.

[9] Barber K.E. Peatlands as scientific archives of past biodiversity. Biodiversity and Conservation (1993);2 474–489.

[10] Maher M., Prasad M., Raviv M. Organic soilless media components. In: Raviv M., Lieth J.H. (eds) Soilless culture: Theory and practice. Oxford: Elsevier; 2008. p459–504.

[11] Raviv M. Production of high-quality composts for horticultural purposes: a mini-review. HortTechnology (2005);15(1) 52-57.

[12] Hoitink H.A.J., Kuter G.A. Effects of composts in growth media on soilborne pathogens. In: The role of organic matter in modern agriculture. Springer Netherlands, 1986. p289-306.

[13] Raviv M., Medina S. Physical characteristics of separated cattle manure compost. Compost Science & Utilization (1997);5(3) 44-47.

[14] Bustamante M.A., Paredes C., Moral R., Agulló E., Pérez-Murcia M.D., Abad M. Composts from distillery wastes as peat substitutes for transplant production. Resources, Conservation and Recycling (2008);52(5) 792-799.

[15] Perez-Murcia M.D., Moral R., Moreno-Caselles J., Perez-Espinosa A., Paredes C. Use of composted sewage sludge in growth media for broccoli. Bioresource Technology (2006) 97:123–130.

[16] Guerrero F., Gasco J.M., Hernandez-Apaolaza L. Use of pine bark and sewage sludge compost as components of substrates for Pinus pinea and Cupressus arizonica production. Journal of Plant Nutrition (2002);25 129–141.

[17] Sanchez-Monedero M.A., Roig A., Cegarra J., Bernal M.P., Noguera P., Abad M. Composts as media constituens for vegetable transplant production. Compost Science & Utilization (2004);12 161–168.

[18] Benito M., Masaguer A., De Antonio R., Moliner A. Use of pruning waste compost as a component in soilless growing media. Bioresource Technology (2005);96 597–603.

[19] Herrera F., Castillo J.E., Chica A.F., López Bellido L. Use of municipal solid waste compost as a growing medium in the nursery production of tomato plants. Bioresource Technology (2008);99 287–296.

[20] Ostos J.C., López-Garrido R., Murillo J.M., López R. Substitution of peat for municipal solid waste- and sewage sludge-based composts in nursery growing media: Effects on growth and nutrition of the native shrub Pistacia lentiscus L. Bioresource Technology (2008);99 1793–1800.

[21] Barile E.R., Sangalang J.B. Variation in sunlight reduction under the canopy of different coconut cultivars and hybrid. Philippine Agriculturist (1990);73 287-295.

[22] Konduru S., Evans M.R., Stamps R.H. Coconut husk and processing effects on chemical and physical properties of coconut coir dust. HortScience (1999):34(1) 88-90.

[23] Stamps R.H., Evans M.R. Growth of *Dieffenbachia maculate* Camille in growing media containing sphagnum peat or coconut coir dust. HortScience (1997);32(5) 844-847.

[24] Evans M.R., Konduru S., Stamps R.H. Source variation in physical and chemical properties of coconut coir dust. HortScience (1996);31(6) 965-967.

[25] Evans M.R., Stamps R.H. Growth of bedding plants in sphagnum peat and coir dust-based substrates. Journal of Environmental Horticulture (1996);14 187-190.

[26] Meerow A.W. Growth of two tropical foliage plants using coir dust as a container medium amendment. HortTechnology (1995);5(3) 237-239.

[27] Meerow A.W. Growth of two subtropical ornamentals using coir (coconut mesocarp pith) as a peat substitute. HortScience (1994);29(12) 1484-1486.

[28] Prasad M. 1996. Physical, chemical and biological properties of coir dust. Acta Horticulturae 450, 21-30

[29] Chan K.Y., Van Zwieten L., Meszaros I., Downie A., Joseph S. Agronomic values of greenwaste biochar as a soil amendment. Australian Journal of Soil Research (2007); 45 629-634.

[30] Asai H., Samson B.K., Stephan H.M., Songyikhangsuthor K., Homma K., Kiyono Y., Inoue Y., Shiraiwa T., Horie T. Biochar amendment techniques for upland rice production in Northern Laos. Soil physical properties, leaf SPAD and grain yield. Field Crops Research (2009);111 81-84.

[31] Glaser B., Lehmann J., Zech W. Ameliorating physical and chemical properties of highly weathered soils in the tropics with charcoal – a review. Biology and Fertility of Soils (2002);35 219- 230.

[32] Lehmann J., Da Silva J.P., Steiner C., Nehls T., Zech W., Glaser B. Nutrient availability and leaching in an archaeological anthrosol and a ferralsol of the Central Amazon basin: fertilizer, manure and charcoal amendments. Plant Soil (2003);249 343-357.

[33] Steiner C., Teixeira W.G., Lehmann J., Nehls T., Luis J., De Macedo L.V., Blum W.E.H., Zech W. Long term effects of manure, charcoal and mineral fertilization on crop production and fertility on a highly weathered central Amazonian upland soil. Plant Soil (2007);291 275-290.

[34] Lehmann J. Bio-energy in the black. Frontiers in Ecology and the Environment (2007); 5 381-387.

[35] Laird D.A. The charcoal vision: a win–win–win scenario for simultaneously producing bioenergy, permanently sequestering carbon, while improving soil and water quality. Agronomy Journal (2008);100 178–181.

[36] Baronti S., Alberti G., Delle Vedove G., Di Gennaro F., Fellet G., Genesio L., Miglietta F., Peressotti A., Vaccari F.P. The biochar option to improve plant yields: first results from some field and pot experiments in Italy. Italian Journal of Agronomy (2010);5 3-11.

[37] Dumroese R.K., Heiskanen J., Englund K., Tervahauta A. Pelleted biochar: chemical and physical properties show potential use as a substrate in container nurseries. Biomass Bioenergy (2011);35 2018–2027.

[38] Altland J.E., Locke J.C. Biochar affects macronutrient leaching from a soilless substrate. HortScience (2012);47(8) 1136-1140.

[39] Tian Y., Sun X., Li S., Wang H., Wang L., Cao J., Zhang L. Biochar made from green waste as peat substitute in growth media for Calathea rotundifola cv. Fasciata. Scientia Horticulturae (2012);143 15–18.

[40] De Boodt M.F., Verdonck O.F., Cappaert I.M. 1974. Method for measuring the water release curve of organic substrates. Acta Horticulturae 37, 2054–2062.

[41] Sonneveld C., Van Den Ende J., Dijk P.A. 1974. Analysis of growing media by means of a 1:1.5 volume extract. Communication in Soil Science and Plant Analysis 5, 183–202.

[42] Fascella G., Zizzo G.V. Efficient propagation technique of *Euphorbia × lomi* Thai hybrids. HortScience (2009);44(2) 495–498.

[43] Prasad M., Maher M.J. 1992. Physical and chemical properties of fractionated peat. Acta Horticulturae 342, 257-264

[44] Verdonck O., Penninck R.D., De Boodt M. 1983. The physical properties of different horticultural substrates. Acta Horticulturae 26, 37-44.

[45] Abad M., Fornes F., Carrión C., Noguera V., Noguera P., Maquieira Á., Puchades R. Physical properties of various coconut coir dusts compared to peat. HortScience (2005);40(7) 2138-2144.

[46] Fascella G. Long-term culture of cut rose plants in perlite-based substrates. In: Zlesak D.C. (ed) Roses. Floriculture and Ornamental Biotechnology 3 (Special Issue 1). Global Science Books; 2009. p111-116.

[47] Fascella G., Agnello S., Maggiore P., Zizzo G., Guarino L. 2010. Effect of controlled irrigation methods using climatic parameters on yield and quality of hydroponic cut roses. Acta Horticulturae 870, 65-72.

[48] Särkkä L.E., Tuomola P., Reinikainen O., Herranen M. 2008. Long-term cultivation of cut gerbera in peat-based growing media. Acta Horticulturae 779, 423-430.

[49] Nowak J.S., Strojny Z. 2004. Changes in physical properties of peat-based substrates during cultivation period of gerbera. Acta Horticulturae 644, 319-323.

[50] Vaughn S.F, Kenar J.A., Thompson A.R., Peterson S.C. Comparison of biochars derived from wood pellets and pelletized wheat straw as replacements for peat in potting substrates. Industrial Crops and Products (2013);51 437– 443.

Chapter 4

Simple Substrate Culture in Arid Lands

Usama Ahmed Aly El-Behairy

Additional information is available at the end of the chapter

http://dx.doi.org/10.5772/59628

1. Introduction

The limited water resources are the major factor that drew the attention towards the use of intensive agriculture in arid land. Protected cultivation was the first step, which started initially at late seventies and intensified at mid-eighties. Maximizing crop yield per square meter of soil as well as per cubic meter of water could be achieved using soilless culture systems [12]. The need for the use of soilless culture is becoming more important in arid lands than several years ago in order to increase the water use efficiency. In addition, continuous cultivation of crops resulted in poor soil fertility, which in turn has reduced the opportunities for natural soil fertility build up by microbes. This situation has led to poor yield and quality. In addition, conventional crop cultivation in soil (conventional open field agriculture) is difficult as it involves large space, large amounts of water and a huge number of labors [1].

Soilless agriculture means growing plants in the absence of land as the normal agricultural, where soil not used as a medium for agriculture but used as a support to the system. This technique can be an alternative to the conventional cultivation in the soil which has problems and difficult. In addition, soilless culture is possibly the most intensive method of crop production in today's agricultural industry in combination with greenhouses.

In protected cultivation, the high levels of crop and continuous cropping inevitably leads to problems of pests and diseases in the soil. The accumulation of these problems leads to a loss of yield and eventually of the crop. Cropping can only continue if some form of sterilization of the soil or rotation of crop can take place. Steam sterilization is not economically viable & not efficient, the use of methyl bromide banned in many European states, and it will band. For production to continue there is inevitably a move towards some form of soilless culture or soil replacement cultivation.

The use of soilless culture has substantially increased during the last decade as it contributes to the intensification of horticultural production and provides high crop yields even in areas with adverse growing conditions [12].

This chapter will discussed the importance of soilless culture of soilless culture in arid land, the constrain of using this type of cultivation, the type of soilless culture used in arid lands and the suitable types of substrate culture for arid lands.

2. Importance of soilless culture in arid land

Soilless culture technique provides a large numbers of advantages could be summarized as follows [9]:

- Standardization of the culture, and of the root environment in particular.
- Excluding soil infection and hence the danger of disinfectant residues.
- Drastic reduction of energy input for the conditioning of the root environment.
- Crop production where the soil is unsuitable.
- Drastic reduction of the water consumption.
- Efficient use of nutrients.
- Better control of vegetative and generative plant development.
- Earlier and higher production.
- Qualitatively better production.
- Rationalization of labor.
- More possibilities for culture mechanization and automatic control.

3. Why we have to move to this type of cultivation in arid lands

Most of the grower in the arid region now a day should move to the soilless culture for the following reasons:

3.1. Increase productivity

The matter of increased yields with the application of soilless culture should 'be examined carefully. It is true that precise control of nutrition to the plants grown in soilless cultures will result in higher yields and quality, but this does not necessarily mean that yields from the best cultures in soil are inferior. Nevertheless, it is difficult to believe that the fast increase in area in soilless culture in the Netherlands and other European countries would have occurred unless commercial growers were confident of some yield increase to help offset the additional cost of soilless culture.

It is of course understandable that if there are soil problems, (i.e. poor soil. saline soil, toxic soil, etc.), then soilless culture will produce much better crops. Many reports published during the last 15 years presenting results on comparison of soilless methods and soil. Most of them show advantages towards the soilless systems, but his was usually been due to a combination of factors such as reduction of labor, higher yields and the greater uniformity of quality due to them or uniform conditions of growth. However, it must mentioned that in many experiments the management of crops in the soil is not controlled properly [9].

3.2. Control of plant nutrition

The accurate control of plant nutrition compared to soilless culture is also one of the most important advantages of soilless culture. This can be seen from: the point of view of the controlled concentrations, which can be applied to the various crops, various environments, stage of plant growth, etc. In addition, harmful elements to plants, above certain concentrations can be kept within safe concentrations (i.e. Mn, B, Zn, Cu, Pb, etc.)

Another important advantage related to plant nutrition in soilless culture, is the uniformity with which nutrient elements can be supplied to the substrate. This is particularly true with water culture and the more sophisticated systems and less true for the aggregate cultures, especially the simplest ones using surface drip irrigation systems (sand culture, etc.).

When using water cultures or aggregate cultures with inert substrates the level of nutrients, supplied to the new crops those chosen by the manager. This is not the case with soil cultures where in many cases excess nutrient levels in the soil from the previous crops produce salinity.

Another advantage of the soilless culture related to plant nutrition is the ability to control the pH and the E.C. of the nutrient solution according to the requirement of the crop and the environmental conditions. Similar control in soil cultures is very difficult and expensive [9].

3.3. Water economy and control

Water is the most important factor for crop production. Protected crops require large amounts of water due to exclusion of rainfall when crop production is required in hot, arid regions of the world; water is like to be a limiting factor not only of availability but also of quality and cost.

The advantage of soilless culture related to the ease of irrigation applies mainly to certain soilless systems, such as NFT and other true hydroponic systems (where the plants have their roots immersed into the nutrient solution) and to sub-irrigated substrate culture, and is not fully applicable to the rest of the soilless cultures using various inorganic or organic substrates. In fact, watering the later, the frequency and duration of irrigation is much more critical substrates with low water holding capacity, compared to soil.

With reference to water saving, certain soilless systems, for instance the closed or recirculated ones, undoubtedly economize water because drainage and evaporation from the surface eliminated by the design and operational scheme of the systems (NFT, "closed" systems, sub-irrigated soilless culture). In addition, with soilless culture, more accurate water supply control practiced.

Furthermore, water culture and sub-irrigated substrate systems save much labor in the time consuming task of checking and cleaning irrigation nozzles. On the contrary, crops grown on substrates and soil, require frequent examination of trippers as these can easily be blocked by calcium carbonate or other compounds especially with a "hard" water supply. The blockage problem can be eliminated either by acidification of nutrient solution or by pretreatment of irrigation water [9].

3.4. Reduction of labor requirement

Out of soil production exclude all cultural practices associated with the cultivation of the soil, sterilization of soil, weed control, etc. Labor requirement for soilless culture is not similar to all soilless systems. Therefore, the system itself, the degree of automation, the type of substrate, the number of crops raised on each substrate, etc. but in any case, generally speaking, there is a saving in labor impute when soilless culture employed [9].

3.5. Sterilization practices

The greenhouse soil must be free from any soil-borne pathogens before the establishment of any new crop. Sterilization is a difficult and costly operation, but necessary and of great importance. It is justified because the greenhouse business require high investment in structures, facilities, plant materials, running costs, etc. and the need to obtain maximum yields and returns, is obvious to have an economically viable operation. The most effective method of soil sterilization is by steaming, but the method is expensive due to the high cost of energy and labor, therefore its application eliminated. Chemical sterilization is less expensive but not without disadvantages, i.e. the use of formaldehyde had the problems of fumes which are highly phytotoxic and the most important chemical, methyl bromide, a very toxic material to handle, has the problem of chemical residues (bromide ions taken up by the crop) and environmental pollution.

It is therefore, of great advantage the cultivation of crops outside of the soil as there is no need for sterilization when materials and substrates used only for one time, because spreading of diseases avoided. When "closed" soilless culture used depending on the system, the need for sterilization varies, i.e. to clean "true hydroponic" culture structures, following the removal of all debris, etc., a dilute rate of formaldehyde used, followed with clean water. In the NFT system, the film that forms the gullies can be replaced. When solid substrates are used, steam or chemical sterilization should be applied if the material is to be used again. In this case, the application of both is easier and economic but in any case, sterilization of soilless culture systems is easier than soil sterilization [9].

http://dx.doi.org/10.5772/59628

3.6. Control of root environment

Possibilities for more accurate control of root temperature, root oxygen supply are more easily to achieve in soilless cultivation[9].

3.7. Multiple crops per year

Due to the absence of the cultivation techniques, operations like soil cultivation, soil sterilization etc., the number of crops per year is increased, in a given production area, because the time interval between crops is nearly short [9].

3.8. Unsuitable soil

Soilless culture offers an ideal crop alternative to soil culture when there is no soil available at all, or there is no suitable soil for crop production, when soil salinity is higher or there are toxic substances into the soil and finally there is an accumulation of soil pathogens into the soil [9].

4. The constraint points for soilless culture in arid lands

- The high temperature most of the year.
- The availability of the soft water.
- The needed water for the cooling.
- The availability for the equipment in the country.
- The availability of the fertilizer for making the nutrient solution in the country.

5. Soilless culture system for producing vegetables in arid lands

Soilless culture divided into three major branches according the root growing media

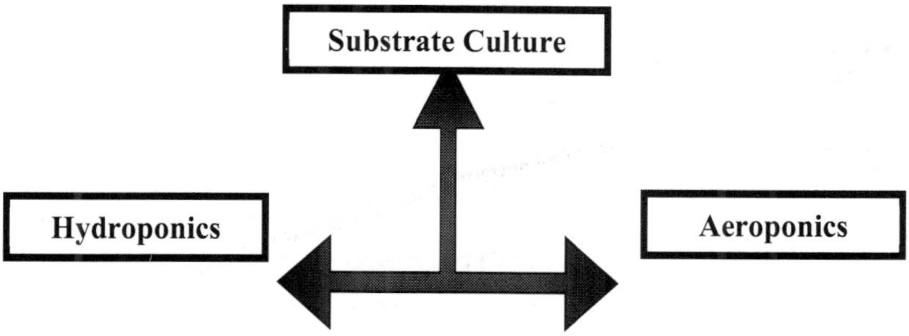

Hydroponics is a technology for growing plants in nutrient solutions (water containing fertilizers). There are different types of hydroponics according the movement of the nutrient solution and the two main systems as follow:

5.1. Deep water culture

The plant roots grow in containers filled with water containing fertilizers (static water). In this system water mixed with nutrient solution and oxygen applied to the plants through nutrient solution using air pump or by mixing water with air. The plants are flouting on the nutrient solution using polystyrene sheet.

5.2. Nutrient Film Technique (NFT)

In this system, the roots of the plants grow in a shallow film of water and nutrient solution inside cultivation channel or tube. The plants inserted in polyethylene (black on white) channel laid on a slope bed. The nutrient solution pumped into the channels in a thin film of nutrient solution and the excess nutrient solution collected and returned to the catchment tank.

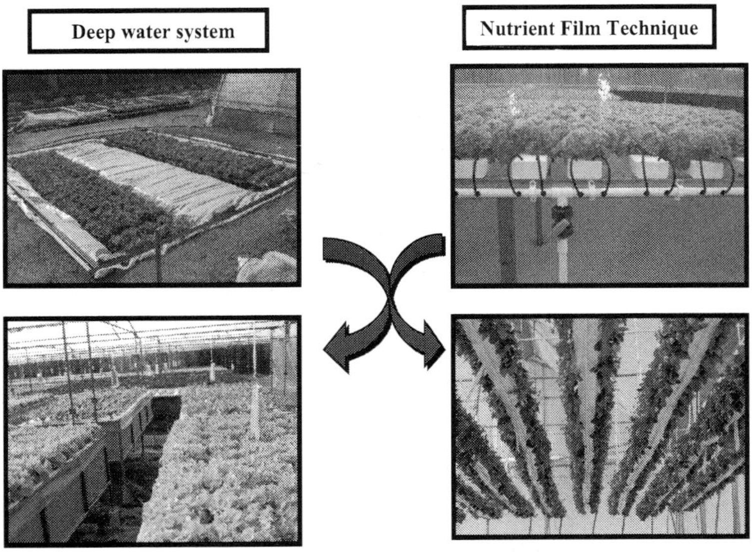

Figure 1. Different systems of deep-water culture and nutrient film technique

5.3. Aeroponics

In this system, nutrient solution sprayed as a fine mist in sealed root chambers. The plants are grown in holes in panels of expanded polystyrene or other material. The plant roots suspended in midair beneath the panel and enclosed in a spraying box. The box sealed so that the roots

http://dx.doi.org/10.5772/59628

are in darkness (to inhibit algal growth) and in saturation humidity. A misting system sprays the nutrient solution over the roots periodically. The system normally turned on for only a few seconds every 2-3 minutes. This is sufficient to keep roots moist and the nutrient solution aerated [1].

Figure 2. Aeroponic system for producing lettuce

5.4. Substrate culture

In this system, a solid medium provides support for the plants. As in liquid systems, the nutrient solution delivered directly to the plant roots. The substrate culture divided according to drainage procedure into two major systems according to drainage procedure

5.5. Open system

The open system is that when the nutrient solution applied to the system with the plants grow and then drained off as waste.

Because the leached or drained solution is not recirculated to the feeder tank, it does not require monitoring and adjustment. Once mixed, it is generally used until depleted. In addition, the quality of the irrigation water is less critical. A content of up to 500 ppm of extraneous salts is easily tolerated, and for some crops (tomatoes, for example) even higher salinities are permissible, although not desirable. It is advisable to monitor the growing medium, particularly if

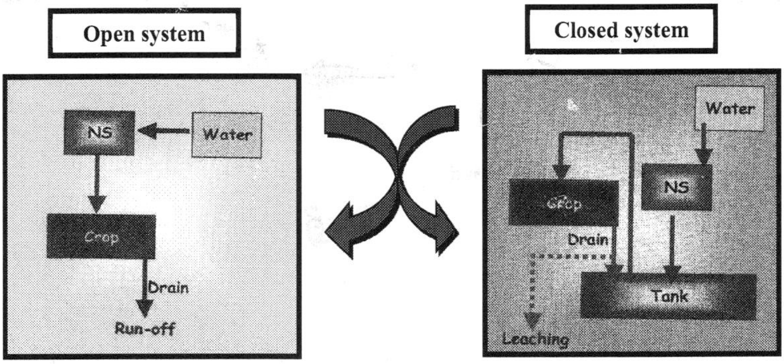

Figure 3. Scheme of the open and closed system for substrate culture systems

the irrigation water is relatively saline or if the operation is located in a warm, high sunlight region. To avoid salt accumulation in the medium, enough irrigation water is used to allow a small amount of drainage or "leaching" from the bags. This drainage should be collected and tested periodically for total dissolved salts [1].

5.6. Closed system

Closed system works in the same way as open system with on important difference that nutrient solution which run-off after each application is collected and recirculated to be used in successive irrigation times.

Closed systems are economical in the use of nutrients, but require frequent monitoring and adjustments of the nutrient solution. Measuring (EC) is a convenient check of total salt concentration, but provides no data on the concentration of major elements, and it is virtually unaffected by the amounts of trace elements present. Periodic chemical analyses are required, usually every two or three weeks for major elements and every four to six weeks for trace elements.

Small farmers commonly practice this regime: Begin with a new solution; at the end of a week add one-half of the original formula to the solution. At the end of the second week, dump the remaining mixture from the tanks or sumps and start all over again [1].

6. Growing media used for growing horticulture crops in the arid lands

6.1. Function of growing media

- Serves as a reservoir for plant nutrients
- Serves as a reservoir for water available for plants
- Must provide gas exchange between roots and the atmosphere outside the root substrate

http://dx.doi.org/10.5772/59628

- Provides support for the plant.

Some individual materials (substrates) can provide all four functions, but not at the required level of each. For example, sand provides excellent support and gas exchange but has insufficient water-and nutrient supplying capacity [9].

6.2. Characteristics of appropriate substrates

- Capacity to hold water

The capacity to hold and drain surplus water depends on the texture of the medium, the size and form of its granules and the permeability. The smaller granules have more surfaces, are close to each other, and therefore, can hold more water. Also the uneven form of granules has a surface area more than granules of even or round form thus the first has higher capacity to hold water.

Consequently, the size of granules should be appropriate so that it can hold proper quantity of water suitable to the crop to be grown [1].

- The substrate should have good aeration and good drainage capacity

The substrate should have a good drainage capacity for draining surplus water and therefore, ensures good aeration around roots. Therefore, we should avoid substrate/medium with fine granules which impedes the movement of oxygen through such granules, reducing the overall aeration condition in the growing environment and leads to asphyxiation of plant roots [1].

- The substrate should be free from harmful or poisonous materials

The substrate should be free from any material, which may cause harm to plant roots or affect plant growth such as sand and small stones of lime origin (contain calcium carbonate). This should be avoided as it can increase the nutrient solution pH to more than 7. This increase leads to sedimentation of iron and phosphorus causing a symptom of deficiency although they exist in the solution.

- The substrate should be supportive to plants growing in it

The substrate should be acting to fix the plants properly. This depends on the texture of the substrate, which should be medium-heavy to fix plant roots [1].

- To be free from diseases incitements

The substrate should be free from different pests and insects so that it would not form a source of infecting plants by different diseases [1].

- To be free from salinity

The substrate should be free from salinity to avoid affecting the growing plants. For instance, the medium made of wood dust usually contains high concentration of sodium chloride due to soaking the wood in salt solution for long periods [1].

- The substrate should be free from weed seeds

This is to avoid being a source of weeds, which will grow and compete with the main crop for nutrition and water. In many cases, weeds would also be a host for some diseases, which would infect and damage the growing plants [1].

- To be slow in the decomposition process

In case of using organic medium, it would preferably be of slow deteriorating nature so that it continues to be in the best condition for the longest period possible. This will reduce the cost of changing the substrate annually [1].

- To be easily transported, handled, and less expensive

There are many kinds of substrates but it is important that the selected one be available in several locations to facilitate its handling and transportation. This would result in reducing the cost of transportation and hence, the preliminary cost of establishing the roof garden. The price of the substrates should be appropriate and acceptable so that the system adopted by all categories of the society [1].

6.3. Types of growing media

Growing media "substrates" can classified as follows:

Inert media: A solid inert material for supporting the plant and provides air and water availability conditions to the roots such as perlite, sand, Rockwool, volcanic gravel, pumice...etc.

Organic media: A natural organic material for supporting the plant such as peat moss, coconut fibers, coco peat, rice hush, wood bark...etc.

There are several raw materials, which used as substrates for roof farming. Such materials differ from one another about its physical characteristics. Due to variations and multiplicity in the forms and types of materials available in the surrounding environment, there is a need for particular criteria, which enables us to select the appropriate material for an agricultural medium (substrate) [9].

6.3.1. Organic substrates

6.3.1.1. Peat moss

The Peat moss considered the most common substrate and widely used at global level. It is a decomposed organic material available in humid locations of the globe called Peat moss mines. This material used separately or mixed with other substrates such as vermiculite or sand.

The peat moss characterized by the following:

- Large capacity to absorb water about 8 folds of its weight at saturation level and drains surplus water.

- Low acidity level.

- High percentage of organic matter (94-99%).
- High porosity (95-98%)

6.3.1.2. Rice husk

The characteristics of rice husk presented below:

- Very light weight.
- Provides necessary aeration for roots of different plants. If mixed with a substrate of bad aeration, it can improve airing and drainage capacity.
- Has a medium capacity to hold water.

6.3.1.3. Coconut fibers

The coconut peat and fibers have recently used as substrate for soil-less agriculture. It obtained from the coconut fruits.

Characteristics of coconut fibers:

- Possibility of use for more than one year without any change in its physical characteristics.
- Its decomposition slow therefore, it would not deteriorate quickly.
- Has capacity to hold water.
- Can provide enough airing to the substrate.

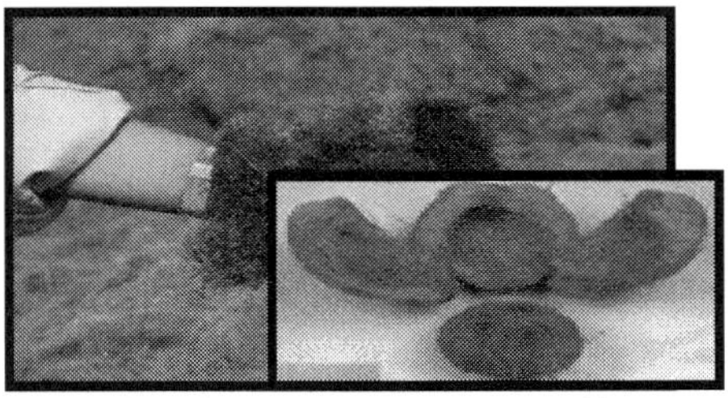

6.3.2. Non-organic substrates

6.3.2.1. Sand

Sand considered one of the best and oldest materials used as a solid substrate for growing plants. It is preferable not to use sands containing lime due to the high rate of calcium carbonate, which acts as a welding material for sand granules and changes the physical characteristics of sand. It is also advisable not to use coast sands due to its high content of salt. It is preferable to use sands of granite or silicone origin as agriculture substrate. The diameter of sand granules is an important factor for successful preparation of agriculture substrate with sand. The course sand cannot hold enough quantity of moisture and very fine sand does not allow a sufficient rate of aeration. Sand characterized by good drainage capacity but its ability to hold water is weak. Therefore, it is preferable to mix it with peat moss or compost.

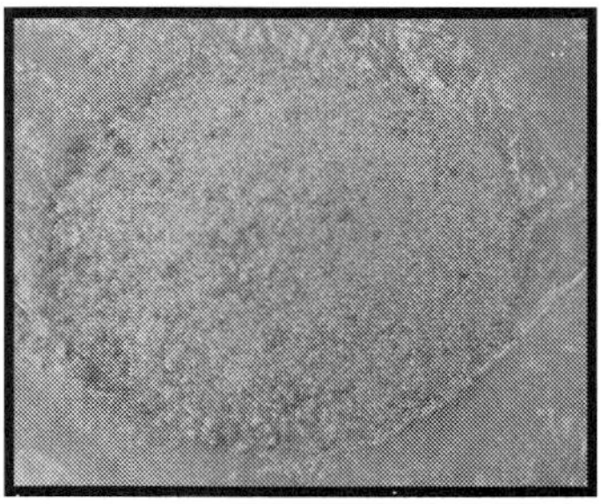

6.3.2.2. Vermiculite

It is dehydrated iron, aluminum and magnesium silicate, which obtained from metallic chips from Mica mines in Africa, Australia, and America. The material to use as a substrate obtained by treating the raw element with a temperature of 1000 centigrade. Therefore, the humidity transforms to vapor which creates an increasing pressure inside its layers, which in turn fragmented to small light pieces of good porosity and characteristics appropriate to soil-less agriculture.

http://dx.doi.org/10.5772/59628

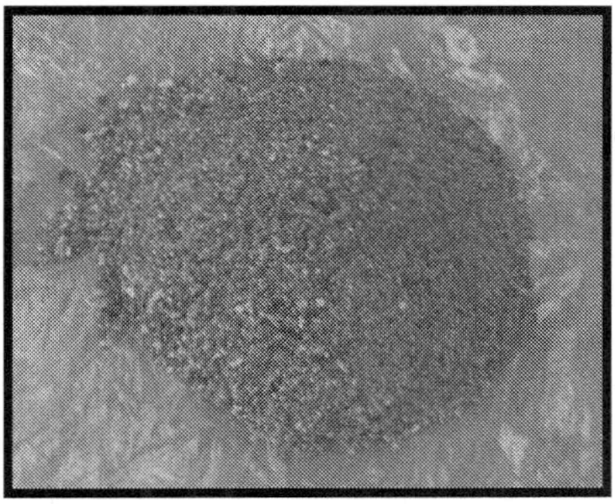

Some characteristics of Vermiculite are:

- High capacity to hold water.
- Contains magnesium and potassium in an absorbable form for the benefit of plants. It has noted that vermiculite is a good water absorption material and therefore, continues to be wet most of time; hence, it is preferable to mix it with other materials to reduce such permanent wet condition so that it would be more appropriate for plant growth.

6.3.2.3. Perlite

This is a volcanic stone originated from volcanic lava of color graded from grey to white and consist of Aluminum Silicate + Sodium and Potassium, which is grinded and heated to high temperature from 900-1000°C. This results to swelling due to exodus of hot air forming air gaps, which cause large expansion, and swelling of granules.

Some characteristics of Perlite are:

- A material of stable physical consistence with no capacity of cationic alternation.
- A light weight material.

- Good drainage capacity while holding enough water. However, irrigation is preferred several times per day to guarantee the water and nutritional elements needed by the plants.
- A substrate of good airing conditions.
- A material of good capillary porosity which facilitates its irrigation by sub-surface method.
- Perlite is widely used either separately with good results or in a mix with other substrates like peat moss to grow several vegetable crops, seeds, flowers, and indoor ornamental plants.

6.3.2.4. Calcined or Expanded clays

Heating montmorrillonitic clay minerals to approximately 690ºC forms calcined clays.

The pottery-like particles formed are six times as heavy as perlite.

Calcined clays have a relatively high cation exchange as well as water holding capacity.

This material is a very durable and useful amendment [11].

Expanded Clay

6.3.2.5. Pumice

Pumice is direct product of acidic volcanism. It is a highly vesicular volcanic glass, silicic in composition and occurs as massive blocks or unconsolidated, fragmented material. The vesicles are glass–walled bubble casts, which give pumice a low density compared to natural glass. Pumicite, the commercial term for fine-grained, fragmented pumice with shards under 2mm in diameter, may be deposited some distance from the source. Pumice is formed from silicic lavas rich in dissolved volatiles, particularly water vapor. On eruption, sudden release of pressure leads to expansion of volatile which, in turn, generates a frothy mass of expelled lava. This mass may solidify on contact with the atmosphere as a vent filling or flow, or may be shattered by a violent eruption. Pumice has many advantages such as high strength-to-weight ratio, insulation and high surface area, which result from the vesicular nature of this rock [11].

http://dx.doi.org/10.5772/59628

Pumice

Some characteristics of pumice

- Pumice a material similar to Perlite from the chemical point except that it contain calcium carbonate which make a problem which react with acid leading to a reduction in the size of the particle. This reduction of particle after using for longer time, the substrate can be compacted.
- It differ than perlite in physical characteristics, where this material is heavier.
- Does not absorb water easily and does not hold it for a long period.
- A substrate of good airing condition.
- Easy to clean and purify [11].

6.3.2.6. Foamy rock

This is a silicon rock of volcanic origin, contains Aluminum, Potassium, Sodium, traces of Iron, Calcium, and Magnesium. The material has several gaps, which formed because of hot vapor exodus before the volcanic lava cools down. This material is available in its natural form and does not need heating but only breaking and grinding to the appropriate size of granules.

Some characteristics of Foamy rock are:

- Foamy rock is a material similar to Perlite from the chemical point of view but differs in physical characteristics, as the first material is heavier.
- Does not absorb water easily and doesn't hold it for a long period.
- A substrate of good airing condition.
- Easy to clean and purify.

6.3.2.7. Rockwool

The use of rockwool has quickly spread in agriculture particularly in Europe where it used to produce many vegetable and ornamental crops.

It is a fiber produced from volcanic rocks and contains Diabase (60%), Lime stone (20%) and Coal (20%).

This mix heated to very high temperature for melting together. The melted material is transformed to fine threads of 5-micron diameter after treatment with fast centrifugal machine and cooling. The threads are then compressed and divided into the required sizes. During the cooling process, the phenol material added to help sticking the rocky wool into a substrate of good porosity.

Important forms and uses of rockwool

- **Germination cubes**: This could be in a single or aggregated form.
- **Seedlings blocks**: To accommodate the small germination cubic's with its contents of plants or the young seedlings directly.
- **Agricultural slices**: To which seeds of proper size transferred, where plant completes its life cycle.
- **Loose (unpacked) rock wool**: This used as substrates for cultivation in pots or mixed with other substrates to improve the characteristics of airing and water holding.

Some characteristics of rockwool are:

- Dry material does not contain any nutritional or non-nutritional solution.
- Sterilized material free of pests, insects, and disease.
- Very light but solid material. This facilitates its preparation and processing.
- A material of high porosity (97% of the total size) which facilitates drainage.
- Facilitates disposal of salt sediments through adding water only in the open system for leaching.

- Easy to sterilize and could be used for more than a year.

The above substrates used separately or in the form of mixture of more than one substrate.

6.4. Substrates mixtures

The above-mentioned substrates could be used in a separate form as agricultural substrate or may be mixed together to attain the best characteristic for the plants to be grown.

The substrate characteristic has a strong impact on success of agricultural operation because it determines the balance between water needed for plant growth and the air necessary for root's breathing. It is therefore, necessary to have small gaps to help holding of water needed for plant life and large gaps required to ensure air necessary for its growth. Some of the most important characteristics, which assessed, are:

- Substrate's weight
- Capacity to hold water
- Acidity (PH)
- Concentration of salts
- Apparent density of substrate
- Cation exchange capacity
- Degree of stability

The above characteristics show the importance of mixing more than one substrates together to achieve the required mixtures. Some of such mixtures were tested and showed good results. Ratios of some mixtures summarized below:

Figure 4. Cucumber grown in sand culture under green house

Substrates	Rate of mixing
Peat moss: Perlite: sand	2:2:1
Peat moss: Perlite	1:1
Peat moss: sand	1:3
Peat moss: sand	3:1
Peat moss: vermiculite	1:3
Peat moss: perlite	1:4

Table 1. Some substrates mixtures. [9]

7. The Recommended substrate system in arid lands

7.1. Sand culture in Egypt

[4] had design a sand culture under plastic tunnels at Dokki Protected Cultivation Center, Cairo.

The design was as follow:

Five 0.8 * 38rn trenches for each tunnel were excavated to a depth of 20 or 40 cm. The bottom of each trench was first leveled and graded to a slope of 12 cm per 40m [5]. The profile of the trench was adjusted to 1 (V) shape. The trench was lined with a water proof polyethylene sheet (200 p) to prevent plants from rooting into the original soil [6]. The surface of the bed was sloped to be parallel to the bottom of the trench.

There inches corrugated perforated plastic pipe placed along the bottom of each trench. The drains were connected at the lower end to a main drain that sink into a 9 m3 nutrient tank. Once the drain pipes are in place, washed coarse sand obtained from Cairo - Alex. Desert road (km 40) filled to a depth of either 20 or 40 cm.

The nutrient tank was 2.0 m length X 1.5 m width X 3.0 m height with 30 cm thick concrete construction, coated with bituminous paint. This tank divided into two equal parts each designed to hold a volume of 30 to 40 % greater than maximum volume required for daily irrigation of each tunnel [10]. A float valve attached to a water refilling line in order to maintain the water level in the tank. The system is designed to recirculate the nutrient solution frequently from the nutrient tank by means of a submersible pump (1 Hp, 220V, and 2 inch in diameter outlet pipe) that was operated by a time clock, one or two times daily.

Some solid particles could be released into the recirculating solution, therefore filtration would be necessary. In fact, the tank acts as a sedimentation tank for the solid particles, which released from the main underground water supply or from recirculated nutrient solution. In addition, two filtration systems were used:

1. A coarse filter "Nylon stocking" was fitted on the outlet of the main drain pipe before the nutrient tank.

2. 150 mesh screen filter were fitted between the circulation pump and the inlet pipe to the main irrigation pipe, in such a way that it is easily removed for cleaning.

The filtration units particularly the screen filter have to be cleaned and replaced fairly frequently because solid particles retained on the screen will progressively reduce the flow rate through the screen.

A drip irrigation system was used with this sand culture [7] with excess nutrient solution (over 50 % of the total applied) to maintain recycling. Such system is termed as a closed system. The drip irrigation system feeds each plant individually by the use of two-liter emitter.

Drip irrigation system of each plastic tunnel contains 50 mm in diameter polyethylene header line. From this header line, 18 mm polyethylene pipe run along each plant row. The emitters were placed in these lateral lies at the base of each plant (50 cm distance between successive plants).

It is worthily to mention that, emitters, pipes fittings of drip irrigation system used for both soil or sand culture and the cover of nutrient tank should be black to prevent algae growth inside the piping system or the nutrient tank.

It is essential that materials used to construct the closed sand culture should not be phytotoxic. In other words, they should not have any harmful effect on the plants. No phytotoxicity has been reported from the use of concrete, bituminous pipes or sheets [5].

7.2. Polystyrene pot system

There are two main systems for the polystyrene pot system:

Figure 5. Vertical pot system (Condensing system)

7.2.1. Vertical pot system (Condensing system):

Different production systems for different crops introduced to small growers in the APRP region by ICARDA-APRP. For production of cash crops such as strawberries and beans, the vertical soilless production system was adapted to maximize growing space by growing the crops vertically. Such technique for strawberry has been investigated for last four years in Bahrain, Kuwait, Oman and Saudi Arabia and currently in Egypt has proved promising from the view point of productivity, cost and water saving. The fundamental structure of the system is the columns, which consist of 8-12 growing containers made from polystyrene on top of each other as seen in the photo. These column of polystyrene pots installed in sloped channel lined by polyethylene sheets to collect the excess of nutrient solution. At the end of the channels there is a PVC tube to collect the return nutrient solution delivering them to filter then to the nutrient solution tank. The column supported by one inch PVC tube from inside the pots. The pots filled substrate (peat moss: perlite 1:4 v/v). The crops are planted in the 4 corners of these containers. The irrigation water and nutrition solution applied to the plants using the drip irrigation and the excess of irrigation recirculated in closed system. The growing containers made locally and the system could be installed in any greenhouse or even in the open field.

Main advantages of the system

The production of strawberries in the vertical hydroponics system was quite successful. The hydroponics system showed followings advantage over the traditional soil-bed production system:

1. 30-50% savings in the cost of the production materials;

2. More yield per unit of water;
3. Double yield per square meter of land area;
4. Longer production season;
5. Increased income due to early season production when prices are high; and
6. Far less incidence of pests and deceases. As a result, lesser chemicals used and higher quality produces obtained.

7.2.2. Simple pot system

For production of cash crops such as tomato, pepper, cantaloupe...etc, the recirculation pot system adapted to maximize growing space by growing the crops in polystyrene pots. Such technique for the cash crops has been investigated for last four years in Bahrain, Kuwait, Oman and Saudi Arabia and currently in Egypt has proved promising from the view point of productivity, cost and water saving. The fundamental structure of the system is simple containers made from polystyrene inserted in a sloped channel lined by polyethylene as seen in the photo. The crops are planted in these containers in substrate consists of perlite: peat moss (4:1 v/v). The irrigation water and nutrition solution applied to the plants using the drip irrigation and the excess of irrigation recirculated in closed system.

Figure 6. Simple pot system cultivated with tomato seedlings

Figure 7. Simple pot system cultivated with cucumber

7.3. Polyethylene containers

There are different types of containers is suitable to substrate culture in arid lands. Also, the different containers shapes create different substrate systems as follow:

7.3.1. Open toped container (Vertical containers)

This type of containers suitable for substrate can hold the water because it can allow a longer column of substrate for the big plants and allowed the water to be drained by gravity.

The idea for these containers to have a holes at the lowermost of the container (5 cm from the bottom) to allow the water to drain. The container filled with small gravel in these 5 cm and then by the chosen substrate. If the open system is used, the containers installed on a bed covered by polyethylene sheets and the drain water is collected and used in another use. If the closed system is used, there a gutter is installed and the containers are installed inside these gutter. The drain water collected and delivered to the nutrient solution after filtering to the nutrient solution tank.

7.3.2. Horizontal bags

The idea of these bags is use for the substrate cannot hold the water like perlite. These bags have a short side and there are holes at the lowermost to drain the excess water. When these bags filled with water, part of these holes will be blocked and keep some water at the bottom of the bags to be a pool can supply the roots with water and nutrient.

http://dx.doi.org/10.5772/59628

Figure 8. Container system using polyethylene bags

Figure 9. Container system using polyethylene pots

Figure 10. Container system using polyethylene containers

Figure 11. Strawberry and sweet pepper cultivated in Horizontal bags

8. Water quality control in the arid lands

The limited water resources and rapid increase in population were the major factors that drew the attention towards the use of substrate in the arid lands.

The best water for substrate cropping is rainwater or water condensed from moisture - laden air. Water from these two sources has virtually no dissolved substances in it. Consequently, there is no build - up of excess ions coming into the substrate installation with the make - up water. An economy in the arid lands, the use of this scarce water can be obtained, if it is mixed with less pure water to provide a blended water in which the concentration of dissolved substances is still acceptable. If the water, that is being used has dissolved in it a substance that is being supplied by the make – up water at a faster than the crop is removing it, then an excess will accumulate in the recirculating solution. If the build - up of the excess is not too rapid, then it is quite realistic to pump out the nutrient solution from the installation after a period of time that was not sufficient duration for an adverse concentration to build up.

[2] suggested to obtain an analysis of the water supply in ppm for the following ions : nitrogen phosphorus,potassium calcium magnesium iron manganese boron, copper, molybdenum, zinc, sodium, chlorides and sulphate. From an inspection of the analytical data it should be possible to decide which ion, or ions, may build up to adversely high concentrations.

Arrangements should then be made for weekly analyses to enable the concentration of the suspect ion (or ions) to be plotted on a graph as the concentration build up. Close observation of the crop will indicate when appearance of the plants begins to be not quite right. However, this method is very - expensive.

On the other hand, [8] suggested a method for deciding when the nutrient solution can be discharging when the hard water was used. This method used successfully in Egypt when the ground water used in the substrate system.

The author suggested that the two most common salts dissolved in the hard water are calcium and magnesium. Electrical conductivity, monitored as EC increases as nutrient salts dissolved in the solution. It follows then that natural salts dissolved in water added to the EC. The EC of the water before nutrients added is known as the base EC. Use the conductivity meter to measure base EC taking care to use a representative sample, i.e., from a pond or other open water source. Collect from "open" water not from puddle edges, from the tap; run the tap for a minute before collecting the sample. If your CF meter is not 'temperature compensated adjust the sample temperature to around 20 "C before taking a reading. The author divided the water to:

1. The EC between 0 to 0.3 m.. mohs follow soft water instructions
2. The EC between 0.4 to 0,8 m.mhos follow hard water instructions
3. The EC over 0.9 m.mohs refer to special adaptations

9. The conductivity program using hard water (Base EC 0.4 – 0.8 m.mhos)

The sum of the effects of using hard water make - up supplies is that, after nutrients additions to a pre-determined level, desired EC subsequent changes in solution EC don't solely reflect the removal of nutrient from the solution by the plants. This situation manifests itself as stable solution EC when make - up water EC additions more or less equate with nutrient losses, or as rising solution EC when make - up water EC additions are greater than nutrient losses. Sometimes, the solution EC may fall if marginally hard water is used. Irrespective of the manifestation, the effect is a gradual decline of the nutrient status of the solution. This decline must be arrested and the following procedure demonstrates how this is done

9.1. At system start - up

Fill the nutrient solution tank with clean water, begin circulation and bring the system to operating capacity. Check, and note, the base CF of the water in the nutrient solution tank. Add the acid to reduce water pH close to desired pH "between 6 - 6.5". The amount of acid used will depend on the hardness of the water. It is useful to keep a record of the amount of acid required so that future treatment of the same volume, after solution discharges, will be rapidly accomplished. Do not overdose; avoid lowering the pH much below 6.0.

Determine desired EC depending on the crop and use the following equation:

Target EC = Desired EC + 1/2 Base EC

Add nutrient stock solution A & B in equal volumes, unless specifically desired to do otherwise, to achieve target EC.

Add nutrient A first and allow this to disperse a little before adding B. Allow time between nutrient additions and monitoring EC for the nutrients to disperse throughout the system. This process may be encouraged by stirring the solution in the circulation tank. When target EC achieved check solution pH and adjust if necessary.

The following examples illustrate this procedure :

- Base EC of make - up water 6
- Solution volume in the nutrient solution tank 210 liters
- Original target EC 1.3
- Make – used water up 70 liters
- Therefore, make - up water volume is 1/3 rd. of solution volume and contributes proportionally to the solution EC i.e., 6 divided by]/3 = 2.
- Target EC = Desired EC + Base EC = 1.3 + '/2 (2) = 1.3 + 0.1= 1.4 m.mhos
- The new target EC = 1.4 m.mhos then add nutrients A & B to bring the solution to the new target EC.

9.2. Discharging the solution

Eventually the target EC will be raised to an unacceptably high value. Generally, this occurrence dictates the time the solution should discharged. The frequency of discharging will regulated by the rate of water removal by the plants of course, but also and particularly by the hardness, manifested as base EC of the make - up water.

The general advice, found to have great practical utility, is to allow the rise in EC to continue until it passes a value 50 % greater than the original EC. When it reaches this value you have to discharge the nutrient solution.

Some mono crop growers a dot the procedure of discharging the solution when the target EC reaches twice the desired EC.

10. The special program for using very hard water (0.9 or above)

The principal problems with very hard water supplies, base EC 0. 9 m.mhos or above, are these:

First, because of the very high level of natural salts present, there is a large oversupply of those which are also nutrient salts, so that a smaller proportion are removed from the system due to take up by the plants.

Second, large amounts of acid are required to neutralize the salts. The result is that' start - up solution EC, already high due to high base EC, is quickly increased by make - up water and acid additions, and there is reduced scope to assign these EC increases to useful nutrients. The practical significance is that the full base EC must be allowed at system start - up when

http://dx.doi.org/10.5772/59628

calculating a target EC and when maintaining nutrient levels in the solution by setting a new target EC. Accordingly, when using very hard water supplies, calculation of target EC uses the following equation:

Target EC = Desired EC + Base EC

Then it follow the same procedure as described in the hard water (EC 0.4 – 0.8 m.mhos)

11. Nutrient solution composition

Plants require 16 essential elements for their growth and development. Without these nutrients, plants cannot complete their life cycles and their roles in plant growth cannot replaced by any other elements. These 16 elements divided into micro and macro element categories as sketched bellow.

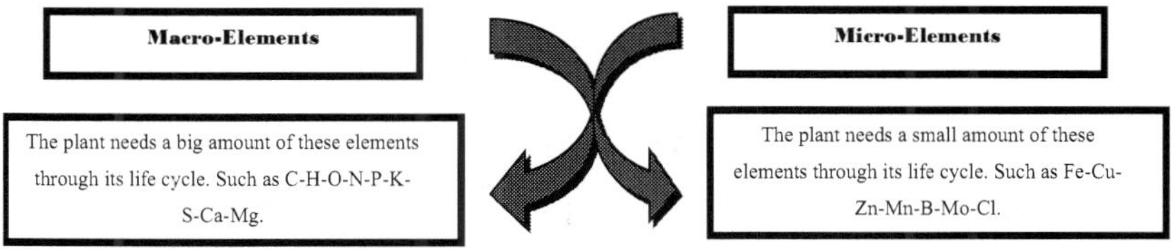

All the nutrient elements required for plant growth have to be present in the nutrient solution for soilless culture systems. Some of the elements may be present in adequate quantities as in the water supply. Other elements will have to be added to the water. These usually are:

N – P – K – Ca – Mg - Fe- Mn – B – Cu – Zn - Mo.

****Ideal Concentrations (ppm) of Elements in Nutrient Solution for tomato is given below as an example**

Elements	Symbol	Concentrations
Nitrogen	N	200
Phosphorus	P	60
Potassium	K	300-350
Calcium	Ca	170
Magnesium	Mg	50
Iron	Fe	3-6
Manganese	Mn	0.5-1.0
Boron	B	0.3
Copper	Cu	0.1
Molybdenum	Mo	0.2
Zinc	Zn	0.05

12. Conclusion

It is clear to solve the problems in the arid lands as shortage of water and lack of technology and the limited income of the grower that using simple but the high technology substrate culture can be a substitute of the soil cultivation. This is because that most of the equipment produced locally, the prices of this equipment's is reasonable and can afforded by the medium class grower special the grower is producing for export. This technique is saving water, there is no need of soil sterilization, there is no for land reclamation.

The most system of the simple substrate system is the column polystyrene pot system for the small plants and the single polystyrene system for the big plants, which can isolate the plants from the high temperature.

Author details

Usama Ahmed Aly El-Behairy

Address all correspondence to: el_behairy2003@hotmail.com

Arid Land Graduate Agricultural Research Institute (ALARI)- Ain Shams University, Cairo, Egypt

References

[1] Abou-Hadid, A. F.; . El-Behairy, U. A; Metwally, Neveen E. and Aly, M. S. 2004. Current situation of soilless culture in Egypt. Regional Training workshop on soilless culture technologies. Izmir, Turkey. March 3-5, 2004.

[2] Cooper, A.J. (1979). "The ABC of NFT", Growers Books, London, 181cp

[3] Cooper, A. (1982). "Nutrient Film Technique", Grower Books, London.

[4] Ibrahim, A., El-Gindy, A., Abou-Hadid, A. and Abdel Ghafar, A. (1989). Recirculated sand culture for tomato under protected cultivation. Egypt. J. Hort. 16, No. 2 pp. 181 - 198.

[5] Jensen. M. H. (1971). The use of polyethylene barriers between soil and growing medium in greenhouse vegetable production. P. 144 - 149. In. Proc. 10 th Natl. Agr. Plastics Conf., Chicago. P. 144 - 149.

[6] Jensen, M.H. and Hicks, N. (1973). Exciting future for sand culture. Amm. Veg. Grower, Nov. (33 - 73).

[7] Johnson, H. (1979). Hydroponics, "A Guide to soilless culture system". Cooperative. Extemion, California Univ.

[8] Molyneux, C.J. (1988). A practical guide to NFT. Nutriculture Ltd.,M awdesley,O rmskirk, Lancashire pp. 153.

[9] Olympios, C. M (2011). Overview of soilless culture: Advantages, constraints and perspectives for its use in Mediterranean countries. CIHEAM Options Mediterranean's Cahiers Options Méditerranéenne vol. 31 – (307-324)

[10] Resh. &M. (1981). "Hydroponic food production; Woodbridge Press Publishing Co., 162.

[11] Verdonck, O. and P. Demeyer, 2004. The infuence of the particle sizes on the physical properties of growing media. Acta Hort., 644: 99-101.

[12] Zayd, A.M., A.F. Abou-Hadid, U.A. El-Behairy and AS. El-Beltagy. (1989). The use of nutrient film technique for the commercial production of greenhouse tomatoes in Egypt. Egypt. J. Hart. 16, No.2, pp. 101 - 110.

Chapter 5

Effect of Different Growing Substrates on Physiological Processes, Productivity and Quality of Tomato in Soilless Culture

Julė Jankauskienė, Aušra Brazaitytė and
Pranas Viškelis

Additional information is available at the end of the chapter

http://dx.doi.org/10.5772/59547

1. Introduction

The most innovative technology of plants growing in greenhouses is growing plants in mineral substrates such as rockwool, vermiculite, perlite, zeolite, ceramsite and others. The origin of substrates is different, some of them are of natural origin while others are produced artificially [1-3]. They also differ in their physical, chemical, and biological properties. Therefore, substrate selection is one of the most important factors affecting plant growth and development in the greenhouse and influencing vegetable quality.

Vegetable-growing rockwool is a widely used substrate for growing of tomatoes and cucumbers under commercial production system. However, one of the biggest disadvantages of this substrate is the need to utilize it. Currently for growing vegetables' different natural substrates are also used and one of these is coconut fiber [4-7]. Substrates of coconut fiber are produced in most countries (like Polland, Netherlands, Belgium, Chezch Republic). Current recycling technologies allow to produce different products in its quality which have its advantages compared to other substrates which are used in greenhouses for growing vegetable [8]. Ready to use coconut fiber substrate may look like dry brick, non-pressed pack as well as blocks' shape. Blocks of coconut fiber are widely used in floriculture, especially for growing roses and gerbera [9,10]. Coconut fiber is a absolutely (100%) organic substrate which is made from recycling the shells of coconuts. It is inert substrate as it does not dissolve upon utilization, size does not change but restrains huge amount of water (more than rockwool). Coconut fiber has other properties such as it is typical to absorb warmth, do not get salty, and it has no pathogens and seeds of weeds [6,11,12]. Substrate of coconut fiber is an alternative for rockwool, no problems appears after utilization. This substrate also has its advantages over rockwool for example; structure of coconut fiber does not change for several years due to its high lignin content. The same structure lasts for 3-4 years. In this way, substrate may be used for few years [8]. Results of last investigations showed that coconut fiber were sufficient substrates for growing of some plants especially for vegetables and grower use these materials as growing media in greenhouse cultures [13]. Albaho and others [14] argued that coconut fiber and its mixes with other substrates could be used as alternative substrate for tomato growing.

Peat and their mixes with perlite, vermiculite, zeolite are the most widely used substrates in greenhouse. In most countries there are analyzed features of zeolite and possibilities to use it for growing of vegetables [15-17]. Zeolites are hydrated crystalline aluminosilicate minerals of natural occurence, structured in rigid third dimension net. This is ecologically clean, inert and non-toxic substance. It is characterized by ion exchange and adsorption features [18,19]. According to Russian scientist, one of the most promising fields of plant-growing is use of natural zeolite as a substrate for seedlings and vegetables to grow [20,21]. There are different reports related to use of zeolite as substrates in hydroponic culture. Technologies of growing cucumbers, tomatoes and green vegetables in zeolite were created in Russia [17,22]. There were also analyzed opportunities how zeolite as a substrate and its mixes with peat could be used in greenhouses [16,19]. It was found that using zeolite less nutrients is missed, efficiency of mineral fertilizer increases. There were analyzed options of using zeolite to grow seddlings of vegetables as well as potted plants [23-25]. Gül and others [26] concluded that the use of zeolite led to increased lettuce plant growth. Most scientific researches reveal the effect of substrates for vegetables productivity [27-29]. Gruda [30] states that it is possible to improve the quality of fruit if suitable substrate is chosen. Other researches show effect of substrates and its mixes for vegetable quality [23,29,31-33].

The aim of this study was to estimate of rockwool and coconut fiber substrates on productivity and quality of tomato hybrids 'Raissa' and 'Admiro'. In addition to establish the optimal amount of zeolite in peat substrate and to evaluate the influence of zeolite-peat mixes on productivity and quality of tomato hybrid 'Ronaldo'.

2. Materials and methods

2.1. Growing conditions

1. The investigations were carried out at the Institute of Horticulture, in the Multi Rovero 640 TR ("Rovero", the Netherlands) greenhouse covered with a double polymer film. The tomatoes were sown at the beginning of February and the seedlings were grown in rockwool growing cubes on the shelvings in a heated nursery and lighted additionally by high pressure sodium lamps (Philips SON-T Agro). At the beginning of March the seedlings were transplanted in the greenhouse (Figure 1). The plant density in the greenhouse was 2.5 plants per m^{-2}. The end of tomato vegetation was the middle of October. Two factors were investigated: factor A – tomato hybrids: a_0 – 'Raissa', a_1 –

'Admiro', factor B – substrate: b_0 – rockwool, b_1 – coconut fiber. Plot area – 8 m^2. Four replications were done in a randomized block design.

2. The investigations were carried out at the Institute of Horticulture, in the Multispan 9.60 SR ("Richel", France) greenhouse covered with a double polymer film. The tomato seedlings were grown in polymer pots filled with peat substrate (Profi 1, Durpeta, Lithuania) (pH 5-6) on the shelvings in a heated nursery and lighted additionally by high pressure sodium lamps (Philips SON-T Agro). In the greenhouse the plants were grown in 25 l peat bags (1 bag – 2 plants) (Figure 2). The plant density was 2.5 plants per m^{-2}. The start of tomato vegetation was the beginning of February and the end was the middle of October. The investigation object was hybrid 'Ronaldo'. Different substrates were investigated: a_0 – peat, a_1 – peat + zeolite (15%), a_2 – peat + zeolite (30%). Plot area – 9.6 m^2. Three replications were done in a randomized block design.

Figure 1. Tomato in coconut fiber substrate

Figure 2. Tomato in peat bags

2.2. Cultivation procedure

In both greenhouses the tomatoes were grown using drip irrigation and fertilized with "Nutrifol" (green, NPK 8-11-35 plus microelements - S, MgO, Mn, B, Zn, Cu, Co, Mo, Fe) (first half of the vegetation), "Nutrifol" (brown, NPK 14-10-25 plus microelements - S, MgO, Mn, B, Zn, Cu, Co, Mo, Fe) (second half of vegetation), magnesium sulphate, calcium and ammonium nitrate fertilizers. There was prepared solution, which was diluted with water in a ratio of 1: 100, and plants were fertilized taking into the account the growth stage (4-15 times a day). Nitric acid was used for water acidification. The concentration of salts in the nutrient solution was EC 2.6–3.0, acidity – pH 5.5–5.8.

2.3. Biometric measurements

During the investigation the plant height was measured at three times during vegetative growth each 10 days after transplanting the seedlings in the greenhouse and the leaves were also counted.

2.4. Determination of photosynthetic pigments and dry matter

For sample preparation of photosynthetic pigment 0.2 g of fresh weight were ground with 0.5 g $CaCO_3$ (Sigma-Aldrich, Germany) and extracted in 100% acetone (Merck, Germany), according to Vetsthtein [34]. Spectrophotometric analysis (spectrophotometer Genesys 6, USA) and quantification of total chlorophylls *a*, *b* and carotenoids were performed at 440.5 nm, 662 nm, and 644 nm wavelengths, respectively. The measurements were performed in four replicates (n=4). The fully formed leaves were analyzed.

To determine dry weight tomato leaves and fruits were dried in a drying oven (Venticell,MBT, Czech Republik) at 105 °C for 24 h. The content of dry matter and photosynthetic pigments in leaves were established at three times during entire growth phase, such as measurement I - at the beginning of flowering, measurement II – at the start of yielding, measurement III – at full yielding.

2.5. Phytomonitoring investigations

The phytomonitoring investigations were carried out on the tomatoes grown in different substrates. The physiological processes of tomato 'Raissa' F_1 were investigated using a phytometric system "LPS-03" created by "PthyTech Ltd."(Figure 3). The following sensors were used for the investigations such as sap water flow, stem diameter evolution, fruit diameter evolution and leaf-air temperature differences (Figure 4). The data of these sensors reflect the plant response to various growing conditions best. In addition, microclimate parameter sensors (those of air temperature and total irradiance) were used. The sensors were fixed according to "PhyTech Ltd." recommendations [35,36]. The sensors of stem diameter evolution, stem flux rate and leaf-air temperature were used as indirect indicators of transpiration. The plants were measured for five days.

Figure 3. The phytometric system LPS-03

Figure 4. Fruit diameter sensor

2.6. Yielding of plants

The tomato yield was recorded at every harvest. Tomato fruits were harvested three times a week, next they were separated into marketable and non-marketable ones. Total yield were calculated by aggregating each harvest.

2.7. Biochemical analysis

The biochemical composition of tomato fruits was investigated at the Laboratory of Biochemistry and Technology, Institute of Horticulture. The following methods were applied in establishing the composition: sugars – by AOAC method [37], carotenoids – spectrophotometrically by Genesys10 UV/VIS spectrophotometer (Thermo Spectronic, Rochester, USA) [38], nitrates – by potentiometrical method using an ion selective electrode [39]. The total soluble solids were determined by a digital refractometer (ATAGO PR-32, Atago Company, Japan). The dry matter content was determined by the air oven method after drying at 105 °C in a Universal Oven ULE 500 (Memmert GmbH+Co. KG, Schwabach, Germany) to a constant weight [40]. Ascorbic acid content was measured by titration with 2.6-dichlorphenolindophenol sodium salt using chloroform for intensely coloured extracts [37]. Titrable acidity was measured by titrating 10 g of pulp that had been homogenised with 100 ml distilled water. The initial pH of the sample was recorded before titration with 0.1 N NaOH to final pH 8.2. The acidity was expressed as the percentage of citric acid equivalent to the quantity of NaOH used for the titration.

2.8. Statistical analysis

The data were analysed by ANOVA statistical package [41]. The Fisher's LSD was used to determine significant treatment effects. Statistical significance was evaluated at $p \leq 0.05$.

3. Results

3.1. Effect of rockwool and coconut fiber substrates on productivity, physiological processes and quality of tomato

During vegetation the tomato hybrids grown in different substrates grew and developed differently. The height and leaf number depended both on the substrate used and on the hybrid itself (Table 1). Tomatoes 'Raissa' F_1 grown in a coconut fiber substrate were 8.1–9.2% higher (insignificant difference) compared with the plants grown in rockwool. Moreover, they had a larger number of leaves. The plants of hybrid 'Admiro' grown in the coconut fiber substrate were slightly lower during the first and the second measures taken (insignificant difference) compared with those grown in rockwool. During the third measure taking the height of this hybrid was equal in both substrates. The 'Admiro' plants grown in rockwool and coconut fiber had the same number of leaves.

Substrate	Measurement I		Measurement II		Measurement III	
	Plant height, cm	Number of leaves, unit	Plant height, cm	Number of leaves, unit	Plant height, cm	Number of leaves, unit
			'Raissa' F_1			
Rockwool	75.0	13.5	98.0	17.0	125.4	21.2
Coconut fiber	81.1	14.4	107.1	18.1	136.9	21.7
LSD_{05}	30.6	1.7	32.6	2.5	33.9	1.5
			'Admiro' F_1			
Rockwool	79.5	14.6	103.4	17.7	130.8	21.2
Coconut fiber	75.6	13.9	102.0	17.5	131.6	21.0
LSD_{05}	17.7	1.1	21.8	1.8	20.7	0.7

Table 1. Effect of substrates on plant height and number of leaves of tomato during vegetation

The content of dry matter in the tomato leaves depended on the substrate (Figure 5). Both hybrids grown in rockwool accumulated a higher content of dry matter in their leaves during vegetation compared with those grown in coconut fiber. The content of dry matter in the leaves of tomatoes 'Raissa' F_1 grown in rockwool was 2.6–8.1% higher (depending on measuring) compared with those grown in the coconut fiber substrate (Figure 5 a). The content of dry matter in the leaves of tomato hybrid 'Admiro' grown in rockwool was higher in all measures taken (Figure 5 b) (insignificant differences).

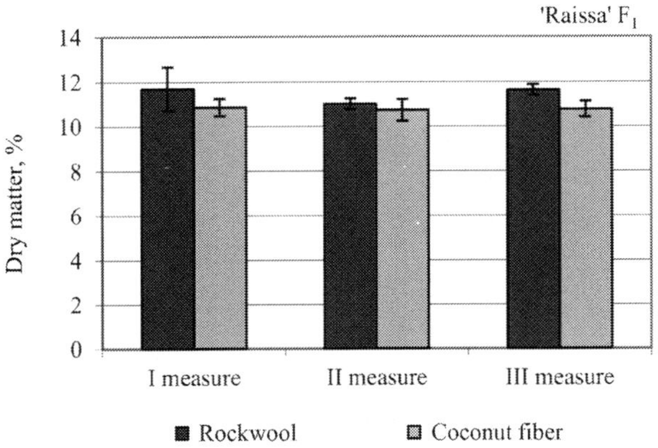

(a)

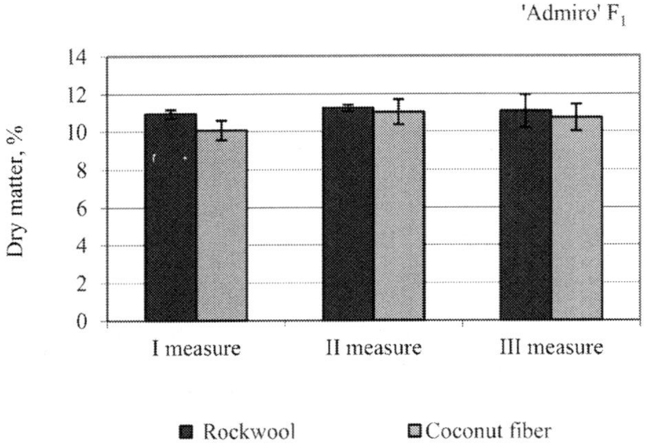

(b)

Figure 5. Effect of substrates on dry matter content in leaves of tomatoes 'Raissa' F_1 (a) and 'Admiro' F_1 (b) during vegetation

The photosynthetic pigment content in the leaves of tomatoes depended on the substrate as well (Table 2). A higher amount of photosynthetic pigments was accumulated in the leaves of both hybrids grown in rockwool. The chlorophyll a + b amount in the leaves of tomato hybrid 'Raissa' was higher by 3.4%; in the case of tomato hybrid 'Admiro' it was 7.0% higher compared with the leaves of the tomatoes grown in the coconut fiber substrate. The chlorophyll a to b ratio in the leaves of both tomato hybrids grown in different substrates was almost similar. The content of carotenoids in the leaves of the tomatoes grown both in rockwool and coconut fiber was more or less the same during vegetation. A slightly lower content was accumulated in the tomatoes grown in the coconut fiber substrate.

Substrate	Photosynthetic pigment content and ratio, mg·g^{-1} fresh mass				
	chlorophyll a	chlorophyll b	chlorophyll a + b	chlorophyll a to b ratio	carotenoids
			'Raissa' F_1		
Rockwool	1.33	0.50	1.82	2.66	0.39
Coconut fiber	1.28	0.48	1.76	2.67	0.37
LSD_{05}	0.21	0.09	0.30	0.03	0.06
			'Admiro' F_1		
Rockwool	1.34	0.50	1.84	2.68	0.39
Coconut fiber	1.24	0.47	1.72	2.64	0.36
LSD_{05}	0.06	0.05	0.11	0.14	0.01

Table 2. Effect of substrates on photosynthetic pigment content and the chlorophyll a to b ratio in leaves of tomato

The intensity of photosynthesis depended on the hybrid of tomato (Figure 6). The photosynthesis intensity of tomato hybrid 'Admiro' was slightly higher compared with those grown in the coconut fiber substrate. The highest intensity of photosynthesis was established with tomato hybrid 'Raissa', when grown in the coconut fiber substrate.

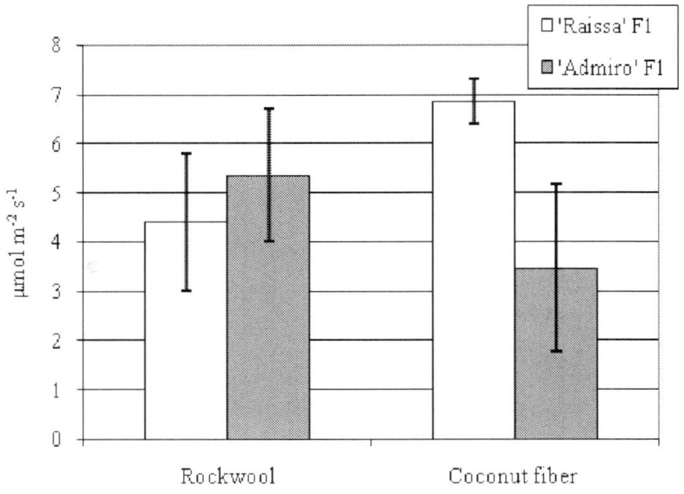

Figure 6. Effect of substrates on photosynthesis intensity of two tomato hybrids viz., 'Admiro' and 'Raissa'

The phytomonitoring investigations were carried out for five days. During the investigations the air temperature within the plant growing zone was about 25 °C and judging from the total irradiance fluctuations the days were overcast with gaps in the clouds (Figure 7). According to the stem flux rate, stem diameter evolution and the difference in leaf-air temperature it is possible to assess the intensity of transpiration and the overall turnover of water in a plant (Figures 8 and 9). The variation of these indicators during 24 hours was similar both in the tomatoes grown in rockwool and in the coconut fiber substrate. In the middle of the day the stem flux rate increased, the stem diameter decreased and the leaf temperature was practically always lower than that of the air. Therefore, it can be proposed that the transpiration in tomatoes was very intensive and a low stem gain per 24 hours indicates that the plants were not supplied with water sufficiently. The tomatoes grown in coconut fiber demonstrated a higher leaf-air temperature difference compared with the tomatoes grown in rockwool. It can be proposed that the transpiration of the latter was less intensive. The more intensive transpiration in the tomatoes grown in coconut fiber had negative influence on fruit growth. Typically, fruits have to grow in a uniform fashion and this substrate practically stopped the daily growth and the growth returned to normal only in the second half of the night (Figure 10). The size increase of the tomatoes grown in rockwool was more uniform. Their growth slowed down in the middle of the day but it returned to normal again in the evening. It can be proposed that the tomatoes grown in coconut fiber substrate demonstrated a higher water demand compared with the tomatoes grown in rockwool.

The coconut fiber substrate had positive effect on the tomato yield (Figure 11). The yield of tomato hybrids 'Raissa' and 'Admiro' grown in coconut fiber was higher compared with those grown in rockwool (insignificant differences). The yield of tomato hybrid 'Admiro' was significantly higher compared with the yield of tomato hybrid 'Raissa' as there were more trusses on the plants formed and the number of fruits in a truss was higher. Somewhat higher early yield was obtained from the tomatoes grown in rockwool. The yield of non-marketable fruits from the tomatoes grown in different substrates was the same: it was 0.24 kg m^{-2} from tomato hybrid 'Raissa' in both substrates and 0.4 kg m^{-2} from tomato hybrid 'Admiro'.

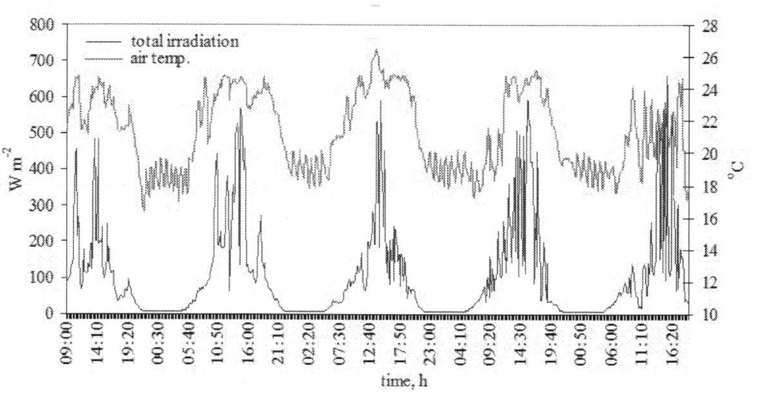

Figure 7. Changes in environmental parameters in greenhouses used for investigating effects of different substrates on the growth, physiological processes and quality of tomato

http://dx.doi.org/10.5772/59547

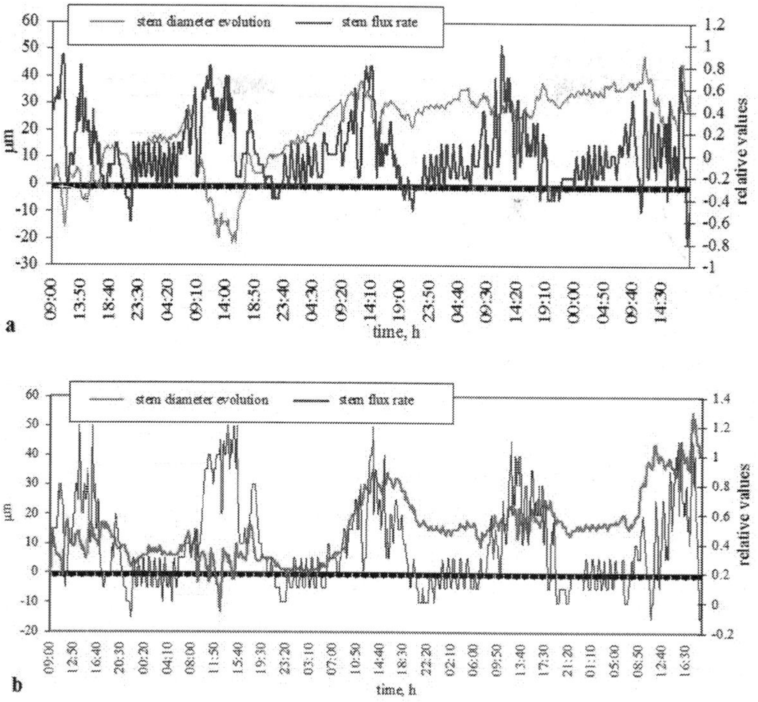

Figure 8. Stem flux rate and stem diameter evolution of tomato hybrid 'Raissa' F_1 grown in rockwool (a) and coconut fiber (b)

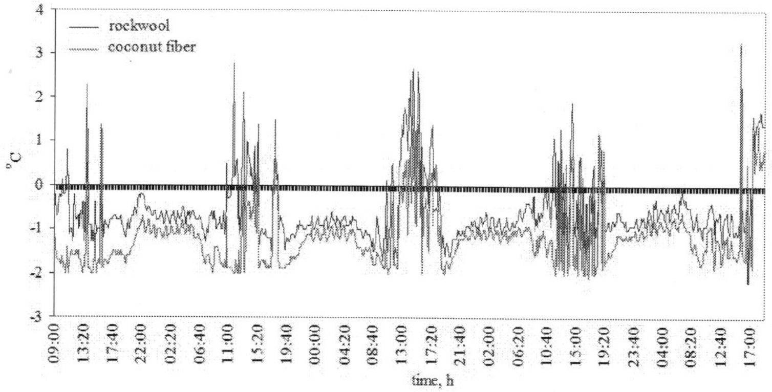

Figure 9. Leaf-air temperature differences of tomato hybrid 'Raissa' F_1 grown in rockwool and coconut fiber substrates

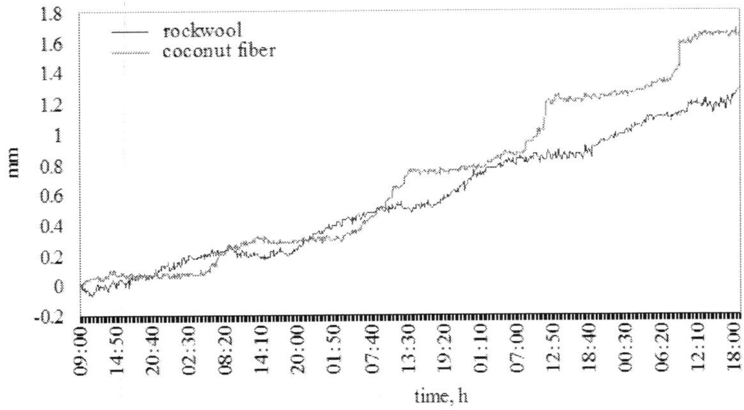

Figure 10. Fruit diameter evolution of tomato hybrid 'Raissa' F_1 grown in rockwool and coconut fiber substrates

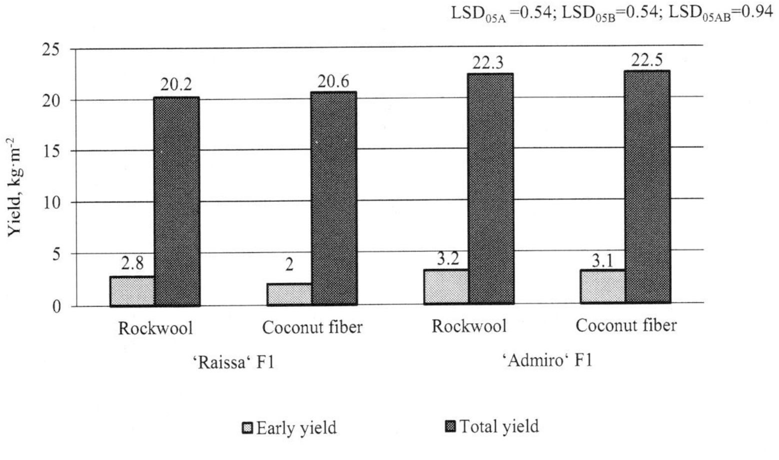

Figure 11. Effect of rockwool and coconut fiber on early and total yield of tomatoes plants grown under greenhouse condition

Tomato hybrid 'Raissa' formed 15 trusses both in rockwool and coconut fiber, however the number of fruits in a truss was different: the number was slightly higher in rockwool compared with coconut fiber (Table 3). Tomato hybrid 'Admiro' formed 15.5 trusses both in rockwool and coconut fiber and the number of fruits was the same. The substrate had no great influence on the average mass of a fruit. The fruit mass of the tomatoes grown in rockwool was slightly higher compared with those grown in coconut fiber. The fruits of tomato hybrid 'Raissa' were somewhat larger – the average mass of a fruit ranged from 137.1 g to 140.1 g, and the average mass of 'Admiro' fruit was between 131.0 g and 135.4 g.

Substrate	Number of fruit in truss, unit	Average fruit mass, g
	'Raissa' F_1	
Rockwool	4.45	140.1
Coconut fiber	4.17	137.1
LSD_{05}	3.18	14.2
	'Admiro' F_1	
Rockwool	4.73	135.4
Coconut fiber	4.93	131.0
LSD_{05}	1.91	10.8

Table 3. Effect of rockwool and coconut fiber on fruit number and average mass of tomatoes plants grown under greenhouse condition

Growing of tomatoes in different substrates had influence on the biochemical composition of fruits (Table 4). Tomato hybrid 'Raissa' grown in rockwool accumulated a higher amount of sugars, dry soluble solids and dry matter (insignificant difference). The amount of ascorbic acid in the fruits of the tomatoes grown in coconut fiber was 1.1 times higher compared with the fruits of the tomatoes grown in rockwool (insignificant difference). Different substrates had influence on the amount of nitrates in tomato fruits: the amount was higher in the tomato fruits grown in rockwool (insignificant difference).

Substrate	Sugar, %			Dry soluble solids, %	Ascorbic acid, mg%	Titratable acidity, %	Carotene, mg%	Dry matter, %	Nitrate, mg kg^{-1}
	inverted	saccharose	total						
				'Raissa' F_1					
Rockwool	3.25	0.13	3.38	4.7	8.8	0.53	2.7	5.4	185
Coconut fiber	2.12	1.14	3.26	4.5	9.5	0.52	2.6	5.0	172
LSD_{05}	1.27	0.32	0.95	1.9	1.3	0.05	4.5	0.6	146.1

Table 4. Effect of rockwool and coconut fiber on biochemical composition of tomato fruit

3.2. Effect of peat and peat-zeolite substrates on productivity and quality of tomato

The admixture of zeolite into a peat substrate had effect on the height of plants. The tomatoes grown in peat-zeolite substrates were lower compared with those grown in peat (Table 5). The tomatoes grown in the peat + zeolite (30%) substrate were 3.1–5.9% lower (depending on measuring) compared with the plants grown in the peat substrate. The tomatoes grown in the peat + zeolite (15%) substrate were the lowest. A lower concentration of zeolite in peat had a greater effect on vegetative plant growth, i. e., the overground mass developed better. An asumption can be made that a greater concentration of zeolite had an effect of better root development but not on the overground plant section.

Substrate	Plant height, cm		
	Measurement I	Measurement II	Measurement III
Peat	49.6	71.0	101.6
Peat + zeolite (15%)	51.1	69.9	100.0
Peat + zeolite (30%)	48.1	67.0	96.9
LSD_{05}	8.4	6.3	6.9

Table 5. Effect of peat and mixture of peat and zeolite substrates on tomato plant height during vegetation

At the beginning of vegetation the amount of dry matter in the leaves of the tomatoes grown in peat was higher compared with those grown in peat and zeolite substrates (Table 6). It was 5.3–8.2% more (Measurement I and II, respectively) compared with the leaves of the tomatoes grown in the peat + zeolite (15%) substrate and 1.1–11.6 % more (Measurement I and II, respectively) compared with the leaves of the tomatoes grown in the peat + zeolite (30%) substrate (insignificant differences). During the Measurement III it was established that the highest amount of dry matter in leaves was accumulated by the tomatoes grown in the peat + zeolite (30%) substrate. The amount was 7.5% higher compared with the leaves of the tomatoes grown in peat and 11.1% higher compared with the leaves of the tomatoes grown in the peat + zeolite (15%) substrate (significant difference). During the entire vegetation the lowest amount of dry matter was accumulated in leaves of the tomatoes grown in the peat + zeolite (15%) substrate.

The content of dry matter in the fruits of the tomatoes grown in different substrates during vegetation was different. The lowest amount of dry matter in the fruits was demonstrated by the tomatoes grown in the peat + zeolite (15 %) substrate. The average data of three measures revealed that the highest amount of dry matter in fruits was accumulated by the tomatoes grown in the peat + zeolite (30 %) substrate and it amounted to 6.4%.

Substrate	Measurement I		Measurement II		Measurement III	
	leaves	fruit	leaves	fruit	leaves	fruit
Peat	11.08	5.95	10.31	5.95	10.06	6.97
Peat + zeolite (15%)	10.24	5.41	9.79	4.49	9.74	6.10
Peat + zeolite (30%)	9.93	5.97	10.20	6.77	10.82	6.47
LSD_{05}	2.23	0.87	2.35	0.40	0.74	0.13

Table 6. Effect of substrates on content of dry matter in leaves and fruits of tomatoes during vegetation

The tomatoes grown in peat and peat + zeolite (30%) substrates accumulated a higher content of chlorophyll compared with the tomatoes grown in the peat + zeolite (15%) substrate (Table 7). The content of chlorophyll was 10.5 %, chlorophyll b – 11.9% and chlorophyll a + b – 10.9% higher compared with the leaves of the tomatoes grown in the peat + zeolite (15%) substrate. The highest chlorophyll a to b ratio was established in the leaves of the tomatoes grown in peat. The carotenoid content was almost the same in the leaves of the tomatoes grown in all substrates.

Substrate	Photosynthetic pigment content and ratio, mg·g^{-1} fresh mass				
	chlorophyll a	chlorophyll b	chlorophyll a + b	chlorophyll a to b ratio	carotenoids
Peat	1.17	0.47	1.63	2.54	0.33
Peat + zeolite (15%)	1.05	0.42	1.47	2.50	0.30
Peat + zeolite (30%)	1.16	0.47	1.63	2.47	0.32
LSD_{05}	0.21	0.07	0.27	0.24	0.07

Table 7. Effect of peat and mixture of peat and zeolite substrates on photosynthetic pigment content and the chlorophyll a to b ratio in leaves of tomato

Zeolite had effect on yield earliness (Figure 12). During the first month of fruiting it ranged from 7.0 kg m^{-2} to 8.0 kg m^{-2} (depending on the substrate). The highest early yield was obtained while growing tomatoes in the peat + zeolite (15%) substrate. It was 11.1% higher compared with the tomatoes grown in peat alone and 14.3% higher than that obtained from the tomatoes grown in the peat + zeolite (30%) substrate (insignificant differences). The total yield was higher in the plants grown in peat and zeolite substrates. The extra yield depended on the amount of zeolite in peat. The yield of the tomatoes grown in the peat + zeolite (15%) substrate was 10.3% (significant difference) higher than that of the tomatoes grown in the peat substrate alone. This effect was related with zeolite's property to accumulate and retain and then release the nutrients to the plants in due time. However, the admixture of higher amounts of zeolite to the substrate had practically no effect on the tomato yield.

The tomatoes grown in different substrates formed the same number of fruits in a truss. In all treatments the number of fruits per truss was between 4.40 and 4.44 u. However, the average fruit mass was slightly different between the treatments and ranged from 133.2 g to 138.1 g (Figure 13). The largest were the fruits of the tomatoes grown in the peat + zeolite (15%) substrate: their mass was 1.9 % higher compared with the tomatoes grown in peat only and 3.7% higher than tomato fruits in the peat + zeolite (30%) substrate (insignificant differences).

The admixture of zeolite into the peat substrate had influence on the biochemical composition of the tomato fruits (Table 8). The fruits of the tomatoes grown in peat-zeolite substrates accumulated less sugars, ascorbic acid and soluble solids. The admixture of zeolite into the peat substrate resulted in a 17.8 -19.6% higher titratable acidity amount (significant difference) compared with the tomatoes grown in peat. In addition, they accumulated a slightly higher amount of carotenoids.

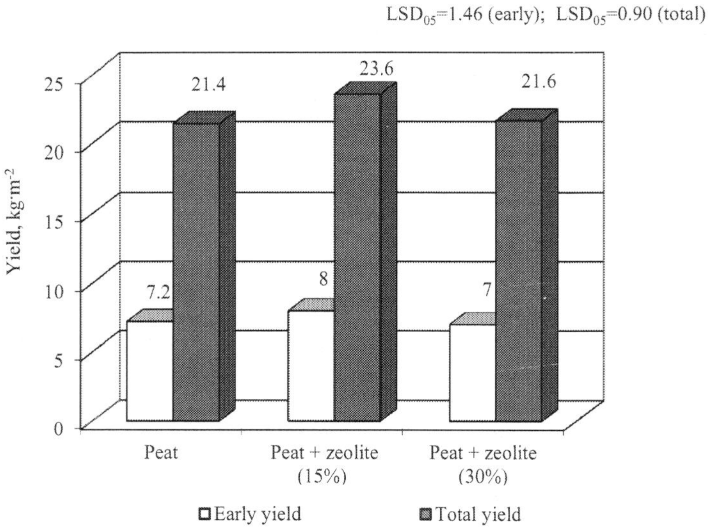

Figure 12. Effect of substrates on early and total yield of tomatoes

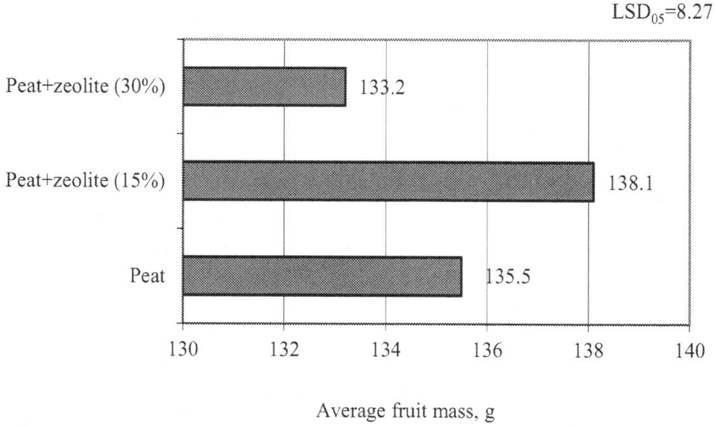

Figure 13. Effect of substrates on average fruit mass of tomatoes

Substrate	Sugar, %	Dry soluble solids, %	Ascorbic acid, mg%	Titratable acidity, %	Carotene, mg%
Peat	5.03	5.2	22.0	0.56	5.5
Peat+zeolite (15%)	4.91	5.1	17.0	0.67	5.7
Peat+zeolite (30%)	4.71	4.9	18.2	0.66	5.6
LSD_{05}	0.98	0.4	1.8	0.02	0.9

Table 8. Effect of peat and mixture of peat and zeolite substrates on biochemical composition of tomato fruits

4. Discussion

4.1. Effect of rockwool and coconut fiber substrates on productivity, physiological processes and quality of tomato

Results of most scientists researches showed that substrates had a significant effect on the plant growth, composition of leaf, total yield and fruit quality [27,31,42,43]. Researchers from different countries analysed the suitability of coco substrates (coconut dust, coco fiber and its mixtures) for growing of vegetables in greenhouses. Data of Lopez et al. [44] showed substrate of coconut dust is proper for growing of tomatoes in greenhouses. It is also characterised as substrate with higher qualities than Canadian peat. Researches of other scientists proved substrates of coconut fiber may be used as organic substrate for growing of plants [4]. In other researches coco substrates was compared to other substrates used in greenhouse vegetable-growing. There was analysed yield of plants while growing vegetables in coco substrate, perlite, rockwool, sawdust, rise husks [12,44-47]]. Coconut fiber was compared to substrate which is produced from waste composting 48]. Alifar et al. [49] investigated the effect of five different growing media for cucumbers' growing. Results showed that the largest stem diameter, the highest biomass were obtained in cocopeat and perlite-cocopeat media. Fecondini and others [50] data reveal hybrid more than substrate where plants were grown in had more influence for phenological observations and biometric parameters of tomatoes. Our data showed the tomatoes grown in rockwool and coconut fiber grow evenly, height of the plants did not differ a lot.

In our research, physiological processes in tomatoes grown in coconut fiber substrate were similar to that were grown in rockwool. Increasing sap water flow, decreased diameter of stem and bigger difference of leaves and air temperatures showed that more intense transpiration was in tomatoes grown in coconut fiber substrate than in plants grown in rockwool (Figure 8 and 9). It may be assumed this had an influence for non-uniform fruit growth in this substrate compared to rockwool (Figure 10). According to physiological researches' data it may be stated that absorption of substrates of coconut fiber and rockwool for water is different.

The content of carotenoid and chlorophyll pigments in the vegetables depend on growing conditions as well as on variety of vegetable [51,52]. According to Islam at al. [53] data there was no difference between content of chlorophyll and dry matter in the leafs of tomatoes grown in coconut fiber and rockwool substrates. There also was no difference on the amount of ascorbic acid in the fruits of tomatoes in these substrates. Our data showed the tomatoes grown in rockwool accumulated more dry matters and pigments of photosynthesis than those grown in coconut fiber substrate.

Various data show substrate has an influence on the yield of tomatoes. Some researches showed yield of tomatoes grown in coconut fiber substrates was higher than grown in other substrates, another researches did not show any difference in the yield [12,48,54]. Kobryn [55] stated bigger yield was got growing tomatoes in rockwool than in substrate Cocovita which is made of coconut palms straw. Jensen's [45] research show there are no fundamental differences between the yield of tomatoes grown in different substrates (perlite, rockwool, coconut, etc.). Carrijo et al. [56] researches state tomatoes grown in coconut substrate were more fertile than those grown in sawdust. Halman [28] data show yield of cherry tomatoes grown in coconut and rockwool was similar. The tomatoes were grown in rockwool, peat, coconut fiber with a different admixture of chips. According to research data, the significantly highest total yield of plants was found in the case of plants grown in peat and in coconut fiber with a higher content of chips in relation to rockwool [57]. Our data showed the yield of tomatoes grown in coconut fiber was little bit higher than those grown in rockwool.

Most scientists found the type of substrate affected the quality of tomato fruit [28,48,58]. Selection of substrate has an influence not only on the yield of plants but on quality of the fruits as well as its beginning of yielding [59]. Growth, yield and fruit quality of tomato grown in coconut fiber were not different from those grown in rockwool [54]. Hallman [28] states the tomatoes grown in cocos substrate had more sugar, acids, there also was less ascorbic acid and licopene. Our data showed growing of tomatoes in rockwool and coconut substrate had an influence on the biochemical composition of fruits (Table 4). Tomato hybrid 'Raissa' grown in rockwool accumulated more sugar, dry soluble solids and dry matter in the fruits but tomatoes grown in coconut fiber accumulated more ascorbic acid. The substrate had no influence on the average mass of a fruit. Our data showed the average mass of fruits was pretty similar growing tomatoes both in coconut fiber and rockwool.

4.2. Effect of peat and peat-zeolite substrates on productivity and quality of tomato

Zeolite and coco substrates may be used in two ways: first, it may be used as a part of mixture of substrates, second, it may be used as only substrate for vegetable growing. There were researches done analyzing growth of tomatoes in mixed perlite and zeolite substrates, mixing both in different ratios [60,61]. According to data, better results were achieved growing tomatoes in zeolite and perlite substrates (ratio 1:1) than growing in cocos and perlite substrates [33]. There also was analyzed the influence of zeolite mixed with other substrates for peppers', lettuce and various flowers yield as well as productivity [15,26,62]. Russian scientists results show lettuce grown in peat and zeolite substrates had smaller content of nitrates, there was bigger yield of cucumbers. Growing vegetable in zeolite substrate there was lower use of fertilizers [20,63]. Our data showed the admixture of zeolite into a peat substrate did not have significant influence for growth of plants. The tomatoes grown in peat and zeolite substrates were a little bit lower than those plants grown in peat.

Seeking for better evaluation of the influence of mixtures of substrates on the plants growth, there was found content of photosynthetic pigments in the leafs of tomatoes. The content of chlorophyll in the leafs of the plant is one of potential productivity indicators. It is often used aiming to find how any of the growing ways or environmental conditions affect the photosynthesis system of the plants. If growing conditions are inadequate, concentration of chlorophylls and ratio of chlorophylls a to b decreases. For the process of photosynthesis chlorophyll a is more important as it reacts to changing conditions of the environment rather [64,65]. Our data found differences between a chlorophyll content in the leafs ot tomatoes grown in different substrates. The smallest content was in the leaves of tomatoes grown in peat + zeolite (15%) substrate. Tomatoes grown in peat + zeolite (30%) substrate and peat only accumulated the same content of chlorophylls. Highest ratio of chlorophylls a to b was in the leaves of tomatoes grown in peat substrate. To sum up, positive impact of zeolite for synthesis of pigments of photosynthesis was not found.

Most researches found mixing of zeolite into other substrate has an influence on the quality of fruits. Aghdak et al. [66] in a study based on the effect of various substrates on qualitative properties of sweet pepper found that addition of zeolite to substrate improves quality of sweet pepper fruits. According to Angelis and other [67] results tomato fruit quality was affected only by tomato variety and not by substrate. Other data showed that no significant differences were found between type of substrate on the amount of total soluble solids, sugars and ascorbic acid in fruits of tomatoes [43,55]. Fecondini et al. [50] data showed hybrid but not the substrate plants were grown in had bigger impact for the yield of tomatoes. Our analysis showed that the content of titratable acidity was bigger in the fruits of tomatoes grown in peat-zeolite substrates.

According to different researchers, the admixture of zeolite into a peat substrate has positive influence on the yield of vegetables [14,68]. Ashraf [69] states, after admixture of zeolite into perlite and pumice not only the features of substrate improves but also the yield of tomatoes increases. Jankauskienė at al. [70] data show, growing of seedlings in peat-zeolite substrate had an influence on the quality of the seedlings though it did not have positive influence on the yield of these vegetables. It is important ratio of zeolite and other substrates, its size of fractions [23,71]. Berar et al. [72] data show after admixturing 25% zeolite into substrate there was found the biggest yield of tomatoes. Živković and others [73] reported admixturing zeolite into peat there was found 35% bigger yield of tomatoes, though the yield in the trials after admixturing 20% and 30% zeolite was the same. Cativelo [23] found more suitable substrate for growing of plants is admixtured substrate with 3-7% of zeolite compared to substrate which has 15% zeolite. Roses were grown in zeolite and perlite substrates mixed in different ratio. The biggest number of roses was picked and its quality was the best when roses were grown in zeolite and perlite substrate when ratio was 25 : 75 [74]. Our data showed smaller amount of zeolite in peat (15%) had bigger impact for yield of tomatoes and average mass of fruit. After admixturing bigger amount of zeolite into peat substrate (30%), the yield of tomatoes was not bigger but in the leafs of tomatoes there were accumulated more dry matters.

5. Conclusions

1. The content of dry matter and photosynthetic pigments in tomato leaves depended on the substrate: the tomatoes grown in rockwool accumulated higher dry matter and chlorophylls in leaves than those of tomatoes grown in coconut fiber substrate. The intensity of photosynthesis depended on the hybrid of tomato. Photosynthesis intensity of tomato hybrid 'Admiro' was more intensive in rockwool and photosynthesis intensity of hybrid 'Raissa' – in coconut fiber substrate. Tomato transpiration was intensive in both substrates, but small stem increase in 24 hours showed insufficient supply of water for plants. This delayed fruit growth in day, especially of these tomatoes, which were grown in coconut fiber substrate. The yield of tomatoes grown in coconut fiber substrate was bigger than this one of tomatoes grown in rockwool. The ascorbic acid content in tomatoes fruit which were grown in coconut fiber was higher than in fruit of tomatoes grown in rockwool.

2. Plants grown in peat-zeolite substrates were lower. The admixture of zeolite into a peat substrate did not influence significantly dry matter accumulation in tomato leaves. The highest chlorophyll a to b ratio was in the leaves of tomatoes grown in peat substrate. Thus, the positive effect the admixtures of zeolite into peat substrate on synthesis of photosynthetic pigments were observed.

   Admixture of zeolite into peat substrate affected the volume of the yield and the average fruit mass. The yield of tomatoes grown in peat + zeolite (15%) substrate was the highest. The tomatoes grown in peat + zeolite (15%) substrate recorded the highest average fruit mass. The tomatoes grown in peat-zeolite substrates accumulated less sugar, ascorbic acid, soluble solids, however, higher amount of titratable acidity and carotenoids.

Author details

Julė Jankauskienė*, Aušra Brazaitytė and Pranas Viškelis

*Address all correspondence to: j.jankauskiene@lsdi.lt

Institute of Horticulture, Lithuanian Research Centre for Agriculture and Forestry, Lithuania

References

[1] Verdonck O, De Vleeschauwer D, De Boodt M. The influence of the substrate to plant growth. Acta Horticulturae 1982;126 251-258.

[2] Olle M, Ngouajio M, Siomos A. Vegetable quality and productivity as influenced by growing medium: a review. Žemdirbystė=Agriculture 2012;99(4) 399-408.

[3] Bhat NR, Suleiman MS, Thomas B, Lekha VS, George P, Ali IS. Growing substrates for organic lettuce production in Kuwait. World Journal of Agricultural Sciences 2013;9(2) 143-147.

[4] Evans MR, Kondoru S, Stamps RH. Source variation in physical and chemical properties of coconut coir dust. HortScience 1996;3 965-967.

[5] Noguera P, Abad M, Noguera V, Puchades R, Maquieira A. Coconut coir waste, a new and viable ecologically-friendly peat substitute. Acta Horticulturae 2000;517 279-286.

[6] Yau PY, Murphy RJ. Biodegraded cocopeat as a horticultural substrate. Acta Horticulturae 2000;517 275-278.

[7] Rahimi Z, Aboutalebi A, Zakerin A. Comparison of different medium for production of Sweet pepper transplant. International Research Journal of Applied and Basic Sciences 2013;4(2) 307-310.

[8] Shishko AA. The use of coconut fiber in greenhouse. Vegetable Growing. Scientific works 2006;12 187-192.

[9] Blom TJ. Coco coir versus granulated rockwool and 'arching' versus traditional harvesting of roses in a recirculating system. Acta Horticulturae 1999; 481 503-510.

[10] Nogurea P, Abad M, Puchades R, Noguera V, Maquieira A, Martinez J. Physical and chemical properties of coir waste and their relation to plant growth. Acta Horticulturae 1997;450 365-373.

[11] Prasad M. Physical, chemical and biological properties of coir dust. Acta Horticulturae 1996;450 21–29.

[12] Carrijo OA, Vidal MC, dos Reis NVB, de Souza RB, Makishima N. Tomato crop production under different substrates and greenhouse models. Horticultura Brasileira. 2004;22(1) 5-9.

[13] Ghehsared AM, Hematian M, Kalbasi M. Comparison of date-palm wastes and perlite as culture substrates on growing indices in greenhouse cucumber. In: The 1th International and the 4th National Congress on Recycling of Organic waste in agriculture 26-27 April 2012, Isfahan, Iran.

[14] Albaho M, Thomas B, Isathali S, George P, Ghloum D. Alternative growing media for growbag tomato production in Kuwait. Acta Horticulturae 2014;1037 1087-1091.

[15] Harland J, Lane S, Price D. Further experiences with recycled zeolite as a substrate for the sweet pepper crop. Acta Horticulturae 1999;481 187–196.

[16] Stamatakis M, Koukouzas N, Vassilatos CH, Kamenou E, Samantouros K. The zeolites from evros region, Northern Greece: a potential use as cultivation substrate in hydroponics. Acta Horticulturae 2001;548 93-104.

[17] Postnikov AV, Zekunov AV, Eliseev NA. Vegetable growing in zeolite. Agricultural Chemistry 1991;11 22-25.

[18] Ming DW, Mumpton FA. Natural zeolites'93: Occurrence, properties, use. International Committee on Natural Zeolites, Brockport, New York;1995.

[19] Mumpton FA. La roca magica: Uses of natural zeolites in agriculture and industry. Proceedings of the National Academy of Sciences of the United States of America 1999;96 3463-3470.

[20] Perfilieva VF. Use of natural zeolites in greenhouses. The use of zeolites from Siberia and the Far East in agriculture 1988 77-80.

[21] Perfilieva VF. Natural zeolites for vegetable growing. Agricultural Chemistry 1991;12 77-79.

[22] Loboda BP, Yakovleva NN. The use of zeolite substrates in greenhouses. Agricultural Sciences. 1999;5 18-19.

[23] Cattivello C. Use of substrates with zeolites for seedling vegetable and pot plant production. Acta Horticulturae 1995;401 251-258.

[24] Manolov I, Antonov D, Stoilov G, Tsareva I, Baev M. Jordanian zeolite tuff as a raw material for the preparation of substrates used for plant growth. Journal Central European of Agriculture 2005;6(4) 485-494.

[25] Demir H, Polat E, Sonmez I, Yilmaz E. Effects of different growing media on seedling quality and nutrient contents in pepper (Capsicum annuum L. Var longum cv. Super Umut F1). Journal of Food Agriculture and Environment 2010;8(3ann 4) 894-897.

[26] Gül A, Eroğul D, Ongun AR. Comparison of the use of zeolite and perlite as substrate for crisp-head lettuce. Scientia Horticulturae 2005;106 464-471.

[27] Peyvast GH, Noorizadeh M, Hamidoghli J, Ramezani-Kharazi P. Effect of four different substrates on growth, yield and some fruit quality parameters of cucumber in bag culture. Acta Horticulturae 2007;742 175-182.

[28] Hallman E, Kobryń J. Yield and quality of cherry tomato (*Lycopersicon Esculentum* Var. *Cerasiforme*) cultivated on rockwool and cocofiber. Acta Horticulturae 2003;614 693-697.

[29] Gao HB, Zhang TJ, Lv GY, Zhang GH, Wu XL, Li JR, Gong BB. Effects of different compound substrates on growth, yield and fruit quality of cucumber. Acta Horticulturae 2010;856 173-180.

[30] Gruda N. Do soilless culture systems have an influence on product quality of vegetables? Journal of Applied Botany and Food Quality 2009;82 141-147.

[31] Padem H, Alan R. The effect of some substrates on yield and chemical composition of pepper under greenhouse conditions. Acta Horticulturae 1994;366 445-451.

[32] Alekseeva TP, Perfilieva VF, Krinitsyn GG. Organomineral fertiliser with prolonged action based on peat. Chemistry of plant raw materials 1999;4 53-59.

[33] Traka-Mavrona E, Gerasopoulos D, Pritsa T, Maloupa E. Growth, fruit yield and quality of tomato in relation to substrate and nutrient source in a soilless culture system. Acta Horticulturae 2001;548 173-180.

[34] Wettstein D. Chlorophyll Letale und der submikroskopishe Formweschsel der Plastiden. Experimental cell research 1957.

[35] Ton Y. Basics of phytomonitoring. Phytomonitoring. 1996;1 3-5.

[36] Ton Y, Nilov N, Kopyt M. Phytomonitoring: a new information technology for improving crop production. Acta Horticulturae 2001;562 257-262.

[37] Helrich K., editor. AOAC: Official methods of analysis.15th ed. Arlington, VA; 1990.

[38] Davies BH. Carotenoids. In: Goodwin TW. (ed.) Chemistry and Biochemistry of Plants Pigments, vol. 1. London, New York: Academic Press; 1996.

[39] Metodiniai nurodymai nitratams nustatyti augalininkystės produkcijoje. Vilnius; 1990.

[40] Food analysis: general techniques, additives, contaminants and composition. Rome: FAO; 1986.

[41] Tarakanovas P, Raudonius S. Agronominių tyrimų duomenų statistinė analizė taikant kompiuterines programas ANOVA, STAT, SPLIT–PLOT iš paketo SELEKCIJA ir IRRISTAT. Akademija; 2003.

[42] Olympious CM. Soiless media under protected cultivation rockwool, peat, perlite and other substrates. Acta Horticulturae 1993;323 215-234.

[43] Peyvast GH, Olfati JA, Kharazi PR, Roudsari ON. Effect of substrate on greenhouse cucumber production in soilless culture. Acta Horticulturae 2010;871 429-436.

[44] Lopez F, Castillo GAB, Torres JLT, Castillo FS. Hydroponic culture of tomato with coconut coir dust as substrate, and its response to ammonium and potassium. Agrociencia 1996;30 495-500.

[45] Jensen MH, Rorabaugh PA, Garcia AM. Comparing five growing media for physical characteristics and tomato yield potential. Proc. of Am. Soc. Plasticulture 1998;27 31-34.

[46] Urrestarazu M, Salas MC, Mazuela P, Ventura F, Castellano D, Sanchez JA. Effect of forced aeration on certain parameters of crop tomato by substrate culture. Acta Horticulturae 2000;659 679-683.

[47] Inden H, Torres A. Comparison of four substrates on the growth and quality of tomatoes. Acta Horticulturae 2004; 644 205-210.

[48] Urrestarazu M, Salas MC, Mazuela P. Methods of correction of vegetable waste compost used as substrate by soilless culture Acta Horticulturae 2003;609 229-233.

[49] Alifar N, Ghehsareh AM, Honarjoo N. The effect of growth media on cucumber yield and its uptake of some nutrient elements in soilless culture. Journal of Science and Technology of Greenhouse Culture 2010;1(1) 19-25.

[50] Fecondini M, Mezzetti M, Orsini F, Gianquinto G, Poppi S. Zeolites in media mixes for soilless production: first results on tomato. Acta Horticulturae 2011;893 1007-1012.

[51] Kobayashi K, Tsurumizu A, Toyoda M, Saito Y. Contents of chlorophyll, b -carotene and pesticide residues in butter head lettuce produced by various cultivation methods. Nippon Shokuhin Kogyo Gakkaishi 1989;36 676-681.

[52] Kimura M, Rodriguez-Amaya DB. Carotenoid composition of hydroponic leafy vegetables. Journal of Agricultural and Food Chemistry 2003;51(9) 2603-2607.

[53] Islam MS, Khan S, Ito T, Maruo T, Shinohara Y. Characterization of the physicochemical properties of environmentally friendly organic substrates in relation to rockwool. Journal of Horticultural Science and Biotechnology 2002;77 143-148.

[54] Shinohara Y, Hata T, Maruo T, Hohjo M, Ito T. Chemical and physical properties of the coconut-fiber substrate and the growth and productivity of tomato (*Lycopersicon Esculentum* Mill.) plants. Acta Horticulturae 1999;481 145-150.

[55] Kobryn J. The effect of substrate type on the yield and quality of tomato fruits (*Lycopersicon Esculentum* Mill.) in glasshouse cultivation. Folia horticulturae 2002;14(1) 53-59.

[56] Carrijo OA, Liz RS, Makishima N. Fiber of green coconut shell as an agricultural substrate. Horticultura Brasileira 2002;20(4) 533-535.

[57] Kleiber T, Markiewicz B, Niewiadomska A. Organic substrates for intensive horticultural cultures: yield and nutrient status of plants, microbiological parameters of substrates. Polish Journal of Environmental Studies 2012;21(5) 1261-1271.

[58] Tzortzakis NG, Economakis CD. Impacts of the substrate medium on tomato yield and fruit quality in soilless cultivation. Horticultural Science 2008;35 83-89.

[59] Lopez J, Vásquez F, Ramos F. Effect of substrate culture on growth, yield and fruit quality of the greenhouse tomato. Acta Horticulturae 2004;659 417-424.

[60] Ymeri A, Gerasopoulos D, Maloupa E. Quality characteristics of 'Daniela' tomatoes grown on a perlite–zeolite culture bag fed with slow release fertiliser. Acta Horticulturae 1999;491 331-336.

[61] Savvas D, Samantouros K, Paralemos D, Vlachakos G, Stamatakis M, Vassilatos C. Yield and nutrient status in the root environment of tomatoes (*Lycopersicon Esculen-*

tum) grown on chemically active and inactive inorganic substrates. Acta Horticulturae 2004;644 377-383.

[62] Mohammad MJ, Karam NS, Al–Lataifeh NK. Response of croton grown in a zeolite–containing substrate to different concentrations of fertilizer solution. Communications in Soil Science and Plant Analysis 2004;35(15-16) 2283-2297.

[63] Makhalov AV. The use of zeolites for growing cucumbers in greenhouses. The use of zeolites from Siberia and the Far East in agriculture 1988 83-86.

[64] Hay RKM., Andrew JW. An introduction to the physiology of crop yield. New York: Jonh Wiley and Sons. Inc.;1989.

[65] Gabryś H, Kacperska A, Kopcewicz J, Lewak S, Starck Z, Strzałka K, Tretyn A. Podstawy fizjologii roślin. Warszawa: Wydawnictwo Naukowe PWN;1998.

[66] Aghdak P, Mobli M, Khoshgoftarmanesh A. Effect of various culture media on apparent and qualitative properties of sweet pepper cv. 'Emili'. 1st National Congress on Hydroponic and Greenhouse Product 2002.

[67] Angelis G, Papadantonakis N, Spano T, Petrakis C. Effect of substrate and genetic variation on fruit quality in greenhouse tomatoes: preliminary results. Acta Horticulturae 2001;548 497-502.

[68] Markovic V, Takac A, Ilin Z. Enriched zeolite as a substrate component in the production of pepper and tomato seedlings. Acta Horticulturae 1995;396 321-328.

[69] Ashraf S. The effect of different substrates on the vegetative, productivity characters and relative absorption of some nutrient elements by the tomato plant. Advances in Environmental Biology 2011;5(10) 3091-3096.

[70] Jankauskienė J, Brazaitytė A. The influence of various substratum on the quality of cucumber seedlings and photosynthesis parameters. Sodininkystė ir daržininkystė 2008; 27(2) 285-294.

[71] Djedidi M, Gerasopoulos D, Maloupa E. The effect of different substrates on the quality of Carmello tomatoes (*Lycopersicon esculentum* Mill.) grown under protection in a hydroponic system. Cahiers Option Mediterranéennes 2003;31 379-383.

[72] Berar V, Poşta Gh. Research concerning the zeolites influence, used in the culture substratum, upon the quality of greenhouse grown tomato. Journal of Horticulture, Forestry and Biotechnology 2011;15(4) 45-47.

[73] Živković C, Kokot J, Mojić S, Rabrenović D. A preliminary study on the effect of the natural clinoptilolite and bentonite additions on the tomato "Nemo Netta" growth. In: Proceedings of the 5th Serbian-Croatian-Slovenian Symposium on Zeolites. May 30th - June 2nd, 2013, Serbia. http://www.zds.org.rs/doc/Proc_148-150.pdf (accesed 8 September 2014).

[74] Maloupa E, Samartzidis C, Couloumbis P, Komninou A. Yield, quality and photosynthetic activity of greenhouse–grown 'Madelon' roses on perlite–zeolite substrate mixtures. Acta Horticulturae 1999;481 97-102.

Chapter 6

Key Irrigation Technologies and Substrate Choice for Soilless Potted Flowers in Greenhouses

Fusheng Ma and Haiyan Fan

Additional information is available at the end of the chapter

http://dx.doi.org/10.5772/59466

1. Introduction

Total area of flower production in China has reached 834, 000 hm^2 In 2009, with the total sales of ¥719.80 billions; the planting area of potted flower has reached 81 710.6 hm^2, with the sales amount of ¥180.80 billions, the amount of protected cultivation area for flower has reached 81767.5 hm^2, where the area of greenhouse is 21 490.5 hm^2 (The website of state forestry administration, 2009). Facility substrate culture has been one of the production models pushed mainly for urban modern agriculture, so it is very significant in practice for the application of soilless culture technology to find an economic, high-yield and efficient culture substrate suitable for the locality. According to statistics, the typical irrigation quota of facility flower is 3 150~3 900 m^3/hm^2 (Beijing Water Science and Technology Institute, 2007) with high quality and clean water resource, but problems such as undefined water consumption amount, low utilization efficiency of irrigation water, lack of irrigation program, wasting of good water and fertilizer resource, unscientific application of water and fertilizer that may impact economic quality of flower are existing.

It is inevitably demand to uncover water consumption of plant for research about natural hydrological cycle, exploration of plant ecological functions and guidance of plant precision irrigation, and assurance of high yield and quality of commercialcrop, thereby this area is always the hot spot of research in various disciplines such as hydrology, meteorology, irrigation and water conservancy, ecotope and so on. At present, there are many findings in water consumption of typical crops (GuoShenghu et al., 2010; ShenXiaojun et al., 2007; Zheng Baoguo et al., 2010; Zhao Ying et al., 2011), fruit trees (Zhao Jinghua et al., 2010; Liu Hongguang et al., 2010; Abrisqueta J M, 2008) and landscape plants (GuHongzhong et al., 2006; Philips N et al., 1996; Sun Huizhen et al., 2004), the research efforts for facility flower water-saving irrigation is still weak, where the research achievement for water consumption concerned soilless culture facility flower is rarely seen, which falls behind rapid development of flower industry seriously.

Research about the irrigation technology for soilless culture flower still lags behind for the walking tube-well irrigation wasting manpower and water-fertilizer resource and then ruining the environment, while the drip irrigation system cannot fit for changing of the density of potted flowers, which hard to manage. But the ebb-and-flow system can recycle water and fertilizer resource as a new irrigation technology. In 1990s, researchers started systematic study on influence to plant growth irrigated by ebb-and-flow technology from fertilizer and substrate (Poole R T et al., 1992; Erin James et al., 2001), comparison of water utilization between various irrigation technologies in a greenhouse (Catherine A. Neal et al., 1992) algae control in an ebb and flow irrigation system (Chase R et al., 1993), some get the patents (Robert W et al.,1994). At the beginning of the 21st century, Chinese researchers started to introduce the ebb-and-flow technology (Ren Jianhua et al., 2004; Yang Tieshun et al., 2009), which settled in Ningxia (GaoYuting et al., 2010), in 2011, a special planting bed was developed (Zhang Xiaowen et al., 2011), in addition to the ebb-and-flow culture test for potted *Hydrangea macrophylla*(Zhang Li et al., 2011), and get many patents of irrigating equipment quickly.

But the ebb-and-flow technology has not been used massively yet for the limitation of its large initial investment, high requirements for the operation and management personnel's technical level and the production facilities after it is introduced to China (Ma Fusheng et al., 2012).The drip arrow technology is used in production of potted plant cultivated with solid substrate as soon as its appearance (Meng Qingguo et al., 2005), the research achievements concentrate mainly on water-fertilizer coupling effect of vegetable, comparison and selection of proper substrate and determination of the lower limit of irrigation (Pang Yun et al., 2006; Yue bin et al., 1998; Wang Ronglian et al., 2009; ZhongGangqiong et al., 2005). Peat soil becomes the main substrate for soilless substrate potted flower, because which can provide available mosisture, fertilizer, air and temperature supplying enviroment. Based on the research results, though various irrigation low limits are key for water utilization efficiency, growth and quality of potted flower (De Boodt M et al., 1983; Zhang Tiejun et al., 2010; Wang Yajun et al., 2003; Zhang Jianhua et al., 2009; Zhang Ning et al., 2011; R. KastenDumroese et al., 2011), there is seldom reports about research results of influence to water utilization efficiency, growth and quality of soilless substrate potted flower from various dripper flows. Scientific basis for selection and matching the drip irrigation emitters of soilless substrate potted flower are lack.

At present, there have been reports concerned influences to growth and development, yield and quality from soilless culture substrate by scholars both here and abroad (De Boodt M et al., 1983; Zhou Yanli et al., 2005; Zhang Tiejun et al., 2010; Wang Yanjun et al., 2003; Zhang Jianhua et al., 2009; Zhang Ning et al., 2011; R. KastenDumroes et al., 2011; David W R. Water et al., 2006; Kang Hongmei et al., 2006; Zhao Jiuzhou et al., 2001; Li Jing et al., 2000; Chen Zhende et al., 1998; Tian Jilin et al., 2003). But the researches are only concentrated on raw material and suitability of substrate based on the indexes of plant growing, yield and quality; in addition to testing evaluation of the indexes such as pore structure, volume-weight and so on for substrate moisture physical property.

Based on the status quo of weak research efforts for high level facility flower water consumption and research lag on new irrigation technology, we take the typical potted foliage flower "Anthurium" as the object of study, the water consumption law and utilization efficiency during the overall process of the facility systematically were studied, in order to provide scientific basis for design of the irrigation system and precision irrigation control of facility potted anthurium. We take drip irrigation as a comparison to study influence of various nutrient solution depths to substrate moisture content, irrigation water utilization, plant growth and water consumption in order to provide scientific basis for the new technology of facility anthurium ebb-and-flow irrigation. For the sake of a proper dripper flow, we choose peat soil (PINDSTURP) that is popular in the market as the culture substrate, study the law of influence to water utilization, growth and quality of soilless potted flower from various dripper flows systematically, in order to provide basis for selection of facility substrate potted flower drippers. At the same time, proceeding from hydrodynamic parameter testing and evaluation of substrate, 8 typical substrates of facility soilless culture were chosen to discuss the water binding capacity, storativity and water availability of them, in order to provide enough theoretical basis for substrate comparison and selection.

2. Experimental site and methods

2.1. Site and experiment description

- Beijing Jidinglida Technology & Trade Co., Ltd.

We start test and research about influence to water utilization and growth of facility drip irritation soilless culture anthurium under various irrigation low limits, facility soilless potted anthurium ebb-and-flow technology, facility potted anthurium proper dripper flow selection at the intelligent multi-span greenhouse of Beijing Jidinglida Technology & Trade Co., Ltd. (39°20′, east longitude 114°20′, elevation 12m).The material of the greenhouse is double hollow polycarbonate sheet with the span of 8m, and the standard width of a room is 4m. There are the temperature and humidity sensors in the rooms, which drive the wetted pads and the fans automatically. The wetted pads and fans are fitted on the south and north walls in the room, in addition to the small fans to accelerate air circulation inside. When the wetted pads and fans are operated, at the same time the door and the skylight are closed, air circulation can be improved and suitable environment is guaranteed. The greenhouse is heated by the heating radiators to keep the temperature between 20°C~30°C. There are external sun louver and internal sunshading to avoid flower leaf burns.

- Central Station for Irrigation Experiment of Beijing

The soilless culture substrate moisture characteristic parameter test and research are performed at the soil physics laboratory of Central Station for Irrigation Experiment of Beijing, Yongledian Town, Tongzhou, Beijing (39°20′, East longitude 114°20′, Elevation 12m).

2.2. Experimental methods

- Water consumption

Weigh the flowerpot everyday with a balance that has a precision of 0.01g. When irrigating, monitor seepage loss collected by the water collector pot by port; weight the pot before and after irrigation, and determine stage water consumption by the water quantity balance method. Weigh the flower sample every other 5~7 days during the test, check the results to guarantee accuracy. Count the water consumption by the water consumption of single pot, and determine the stage water consumption with the water quantity balance method, refer to formula (1).

$$ET_i = W_1 - W_2 + I - D \quad (1)$$

Where, ET_i is the water consumption at period I; W_1 is the flowerpot weight at the beginning of the period, g. Pot^{-1}; W_2 is the flowerpot weight at the end of the period, g. Pot^{-1}; I is the flowerpot irrigation during the period, g. Pot^{-1}; and D is the flowerport water leakage during the period, g. Pot^{-1}.

- Crown diameter and Plant height

Measure the crown diameter of the sample in the directions of east-west and north-south with a ruler, and take the mean values in both directions as the representative values of the sample crown diameter. Take the pot edge as the datum, measure the distance from the pot edge to the peak of the sample as the representative value of the plant height. Take the distance between the spathe to the root of the petiole as the length, and the distance of the widest as the width.

- Substrate moisture physical property

Measure the dry volume mass and the substrate water holding capacity of the cultural medium with the conventional methods (Jiang Shengde et al., 2006).

- Substrate moisture content

The dry substrate mass is determined by the dry unit weight filling before testing, the empty pot mass is determined by weighing before testing, and the overall mass of the flower and the pot is the average mass of the 5 sample pots. Take 3 plants from the same treatment in protecting line, which are same as the sample plant weighted, weigh their net mass and take the mean value as the representative value of plant mass, the testing period is about 5 days. Take the testing day as the middle, and 2 days both before and after it to use the mass value of the same plant, and calculate the moisture content of the substrate day by day. Subtract the plant mass, the dry substrate mass and the pot mass from the overall mass of the testing flower and the pot to get the water mass in the substrate. Divide the mass with water by the unit weight of water, and then by the substrate volume to get the moisture content of substrate.

- Substrate porosity

The saturated gravity drainage method is used to determine the porosity (Jiang Shengde et al., 2006), which including determination of total porosity, the water holding porosity, the aeration porosity and the gas water ratio. Weigh a vessel with a known volume (V) and weigh its mass (W_1), fill the testing substrate according to the design unit weight and weigh its mass (W_2). Soak the vessel with substrate for 24 hours and weigh its mass (W_3), encase the upper port of the vessel by wet gauze with a known mass (W_4) to avoid seepage of fine particle. There shall be no water seepage when inversing the vessel, and weigh its mass (W_5). The formula for the porosity of the substrate is:

$$\varphi = \left[W_3 - W_1 - \left(W_2 - W_1\right)\right] / V \times 100\% \tag{2}$$

$$\varphi_t = \left[W_3 + W_4 - W_5\right] / V \times 100\% \tag{3}$$

$$\varphi_c = \left[W_5 - W_2 - W_4\right] / V \times 100\% \tag{4}$$

$$d = \varphi_c / \varphi_t \tag{5}$$

Where φ = Total porosity, %; φ_t = Aeration porosity, %; φ_c = Water holding porosity, %; d = Gas water ratio; the units of W_1~W_5 are g; and the unit of V is cm^3.

- Substrate permeability coefficient

It is determined with the constant head method (Tian Jilin, 2003). Fill the cutting ring as required by the test, soak it for 24h before take off. Connect an empty cutting ring on it, seal the connecting place to avoid leakage. Add filter paper at the bottom of the connecting cutting ring. Add water into the empty cutting ring till the water level 1mm lower than the edge of the cutting ring. Time when the funnel starting dripping, record the seepages at 1, 3, 5, 7, 10, 15, 20min..., at the same time, add water into the empty cutting ring to the initial level and record the water temperature. End the test when the seepage becomes stable. Repeat each treatment for 3 times and average them. Hereinafter is the formula for the permeability coefficient:

The permeability coefficient at t°C

$$K_t = vL / \left(h + L\right) \tag{6}$$

Where v — Permeability speed, mm min^{-1}; h — Height of water layer, cm; L — Height of substrate, cm

The permeability coefficient at 10°C

$$K_{10} = K_t / (0.7 + 0.03t) \quad (7)$$

- Substrate moisture characteristic curve

Determine the moisture contents of substrate under 0, 0.01, 0.03, 0.05, 0.1, 0.3, 0.5, 1 and1.5Mpa respectively with the 1500F1membrane pressure gauge and dray the moisture characteristic curve.

- Substrate horizontal diffusivity

Determine it with the horizontal soil columnInfiltration method (Lei Zhidong, 1988). Put the substrate (Or soil) with the air drying unit weight as required by the test into the organic glass tube with the diameter of 5cm and the length of 50cm, ignore the gravity, water it with the Markov bottle. Refer to Fig. 1 for the test unit. Record the variation of water level in the bottle and the time when the wetting front passes every 1cm. At the end of the test, take soil (Or substrate) from the wetting front quickly, weigh it and dry it by heating, so get the moisture content distribution of the soil (Or substrate) column. With the testing time t and the moisture content distribution of the soil (Or substrate) column at that time, λ corresponding to various θ can be calculated with the formula $\lambda = xt^{-1/2}$, so the formula for the horizontal diffusivity is:

$$D(\theta) = -\frac{1}{2}\frac{\Delta\lambda}{\Delta\theta}\sum_{\theta_0}^{\theta}\lambda\Delta\theta \quad (8)$$

Where $D(\theta)$ — Horizontal diffusivity, $cm^2\ min^{-1}$; θ_0 — Initial moisture content, $cm^3\ cm^{-3}$; θ -- Moisture content, $cm^3\ cm^{-3}$; λ —Boltzmann transformation parameter, $\lambda = xt^{-1/2}$; x — The moving distance of the wetting front at t, cm; t— Time, min

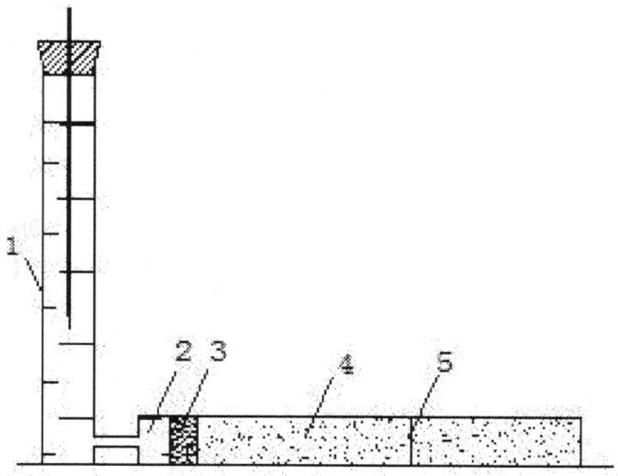

Figure 1. Experimental equipment of measuring soil water diffusivity

1. Markov bottle 2.Water chamber 3. Filter layer 4. Horizontal soil column 5. Wetting front

- Evaluation method

The matrix method is used for substrate comprehensive evaluation. A matrix method refers to form a matrix with all development activities and all environmental factors impacted, establish direct causality, in order to show the impact of the activity on the environmental factor (Duan Zhaolin, 2003). In this paper, the values in the order of merit of 8 substrates under influence from the total porosity, gas water ratio, permeability coefficient, water availability and transmissibility are formed, the better, the larger. Finally, perform comprehensive evaluation to various substrates according to the synthesis score.

3. Impact of various low irrigation threshold on water utilization and growth of facility drip soilless culture anthurium

3.1. Experimental plant and design

The anthurium is chosen as the testing object, its variety is "Flame", which belongs to *Araceae* and *Anthurium*. The culture substrate is mixture of PINDSTURP peat soil imported from Denmark and perlite from China in a volume ratio of 10:1, its dry volume mass is 0.16 g/cm^3, and the volume water holding rate (θ_{FC}) is 0.4315. The tested anthurium is cultivated in a pot, the top diameter of the pot is 16cm, the bottom diameter is 10cm, the height is 12.8cm, the volume is about 1.4L, and the substrate is filled about 2cm to the pot top. Refer to Table 1 for testing treatment.

Testing stage	Lower limit of irrigation/(%)				
	T1	T2	T3	T4	T5
Whole growing period	90	80	70	60	50

Note: The lower limit of irrigation is counted in percentage of the substrate water holding capacity.

Table 1. Experimentle treatments

In this paper, all treatment irrigation quota are controlled by the substrate water holding capacity as the upper limit. The test starts from July 4, 2009 when transplanting to November 5, 2009, takes 124d. The placing density of tested plantlet is 15 pot/m^{2}in the seeding stage, and 10 pot/m^2 in the later stage. The plantlet is 18cm in height with uniform growth when transplanting. Use the pressure compensating dripper with water yield of 3.85 L/h (The working pressure range is 0.05~3 MPa), with the flat perforated pipe configured 4 outlets. Connect the drip arrow with uniform 1m long capillary tube, insert shallowly in the pot substrate from the side direction about 1cm within the center area of the root system. Determine the irrigation quota with the substrate water holding rate as the upper limit according to the relative lower control limit. Place the pot on the bracket, and put the PVC collector under the bracket to collect water leakage. Repeat 5 times per treatment. There are the protection lines at both sides of the monitored sample plant. The self-prime stabilized pressure pump with the pressure tank is used; the hydraulicproportioning fertilization pump made by USA is used for fertilization with water.All test plots are distributed randomly, the treatment measures such as farming, fertilizing, and pest control are same.

3.2. Results and discussion

3.2.1. Impact of various low irrigation threshold on Moisture content of substrate with drip irrigation

Refer to Fig. 2 for influence to moisture content of substrate from various low irrigation threshold. At the beginning of the test, adjust the basic moisture contents of all treatment substrates to the water holding capacity of the substrate. There is no irrigation for all treatments from July 4 to 7, so the moisture contents of substrate of various treatments have no large difference as shown on Fig. 1. From July 7 to 12, T1, T2 and T3 reach the lower irrigation limit and start to irrigate, the moisture contents of the 3 treatment substrates go down in turn from T1 to T3, but all of them are higher than T4 and T5 that do not irrigate and have not very large difference in moisture contents. After July 20, the overall performance of the moisture contents of substrate are 82%~96%θ_{FC} of T1, 75%~94%θ_{FC} of T2, 64%~88%θ_{FC} of T3, 56%~80%θ_{FC} of T4 and 44%~76%θ_{FC} of T5, which goes down from T1 to T5 along with decreasing of the low irrigation threshold. It identifies with the research result of influence to the soil moisture content from the lower irrigation limit (Karam K. et al, 2011).

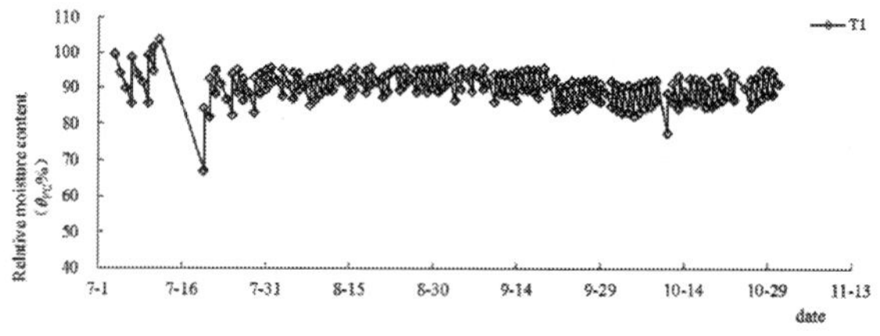

a. T1 (The lower irrigation limit is 90% of the water holding rate of substrate)

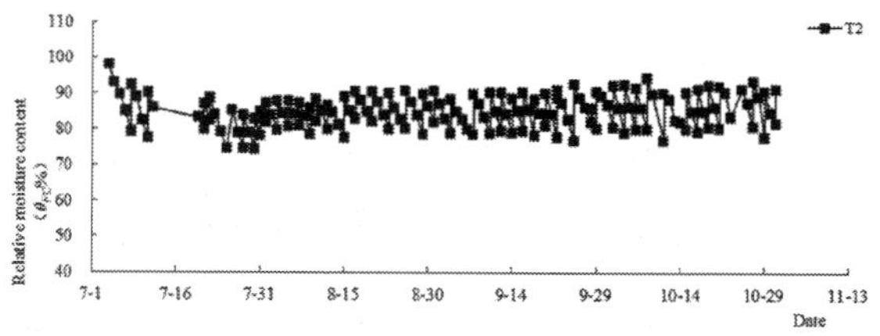

b. T2 (The lower irrigation limit is 80% of the water holding rate of substrate)

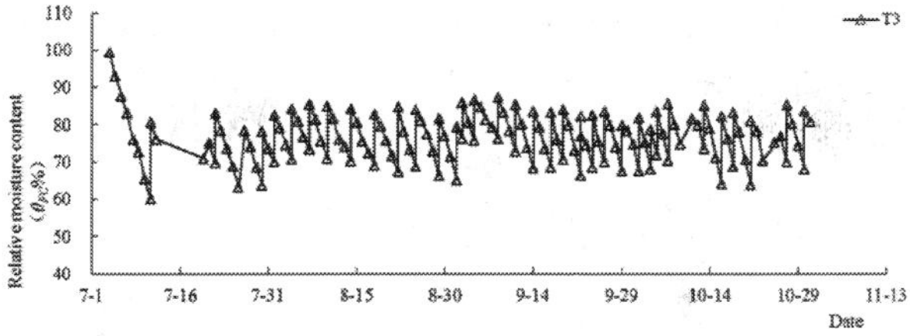

c. T3 (The lower irrigation limit is 70% of the water holding rate of substrate)

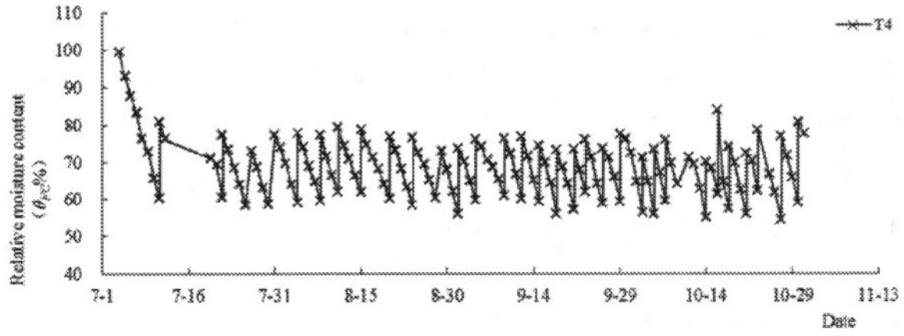

d. T4 (The lower irrigation limit is 60% of the water holding rate of substrate)

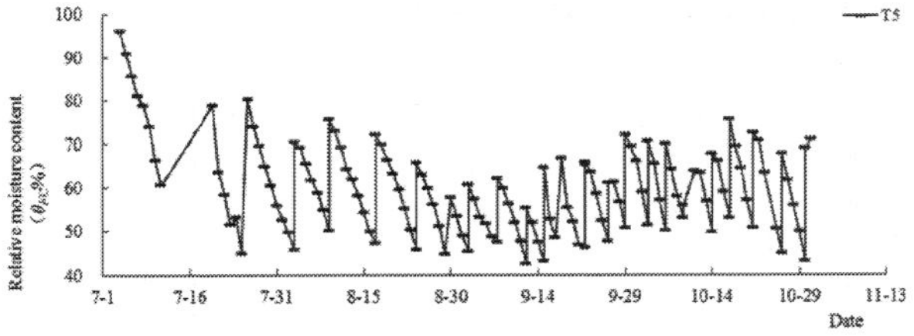

e. T5 (The lower irrigation limit is 50% of the water holding rate of substrate)

Note: The relative moisture contentis counted in percentage of the substrate water holding capacity.

Figure 2. The influence to moisture content of substrate from various low irrigation threshold

Refer to Fig. 3 for the influence to the substrate moister content from various low irrigation threshold after irrigation, while the influence to the field water utilization coefficient is shown in Fig. 4.

As shown in Fig. 3, the substrate moisture contents after irrigation go down along with decreasing of the low irrigation threshold. During the test, the mean values of T1, T2, T3, T4 and T5 after irrigation are 93.32%θ_{FC}, 89.31%θ_{FC}, 83.31%θ_{FC}, 76.02%θ_{FC} and 66.95%θ_{FC} respectively, all of them are lower than the level of substrate water holding capacity measured with conventional methods. It is caused by the substrate drip irrigation local water supply and the physical property of organic medium changing along with the substrate moisture content. Based on the measured results during the test, the average water leakage losses of all irrigations

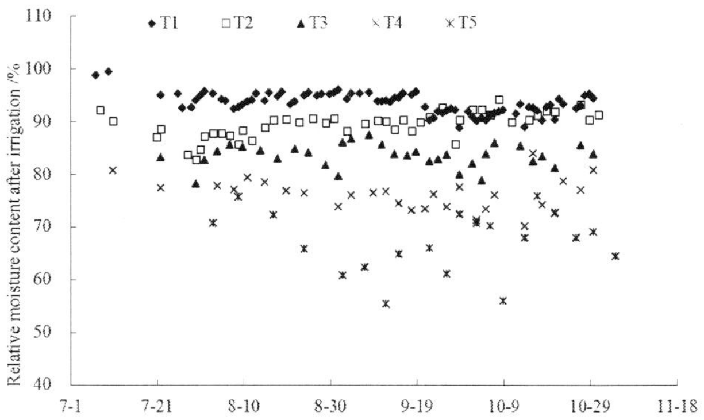

Note: The relative moisture content is counted in percentage of the substrate water holding rate.

Figure 3. The influence to the substrate moister content from various low irrigation threshold after irrigation

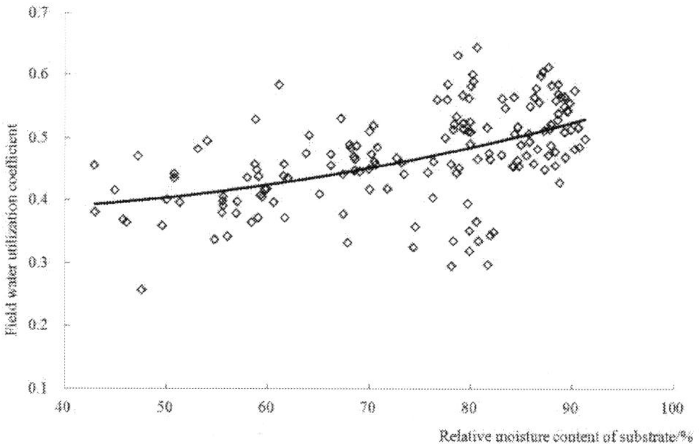

Note: The relative moisture content is counted in percentage of the substrate water holding rate.

Figure 4. Influence to the field water utilization coefficient from various low irrigation threshold after irrigation

for T1, T2, T3, T4 and T5 are 22.30, 49.05, 78.02, 107.74 and 138.62g/pot respectively, the irrigation leakage losses increase along with decreasing of the low irrigation threshold.

The field water utilization coefficient is the ratio between the available water irrigated into field (For dry farmland, it refers to the irrigation water stored in the planned moisture layer) and water discharge from the last stage of the fixed ditch (Field ditch) (Guo Yuanyu et al., 2006). In this paper, it refers to the ratio between the water stored and absorbed by the substrate in the port and the total water irrigated in the port. The total water irrigated in the port can be measured, and the water stored by the substrate is the difference between the total water and leaked water. Refer to Fig. 4,under the test conditions, the total field water utilization coefficient goes down along with the substrate moisture content before irrigation. During the test, the average field water utilization coefficients of T1, T2, T3, T4 and T5 are 0.519, 0.471, 0.439, 0.419 and 0.402. While the irrigation water utilization coefficient specified in *Technical code for microirrigation engineering* (GB/T 50485-2009) shall not be less than 0.9, so the field water utilization coefficient in the tested substrate is even less than it.

In conclusion, the lower irrigation limit is the key factor for the substrate moisture and field irrigation water utilization coefficient (Both of them go down along with the lower irrigation limit) under the conditions that may affect water supply of drip irrigation. Based on their research about influence to substrate physicochemical properties from various water supply modes, Qi Haiying et al. (Qi Haiying et al., 2009) find that various water supply modes can affect the substrate pore structure, substrate volume, salinity, and acid and alkali environmentdramatically. So the change feature of the substrate moisture content and the field irrigation water utilization coefficient maybe the influence result to the tested culture substrate volume and pore structure from local water supply of the drip arrow, while the water absorption, storage and dissipation mechanism of soilless culture substrate needs more intensive study.

3.2.2. *Influence to anthurium irrigation water amount from various low irrigation threshold treatments under drip irrigation conditions*

Refer to Table 2 for influence to anthurium irrigation water capacity from various low irrigation threshold under drip irrigation conditions.

Treatment	Irrigation quota/ $(g \cdot pot^{-1})$	Irrigation period/d	Irrigation quota/ $(g \cdot pot^{-1})$
T1	46.36	All are 1~2 d during the test.	4865.4
T2	92.72	All are 2~3 d during the test.	5124.1
T3	139.08	3~4 d before 60 d after planting, and 2~3 d from 60 to 120d.	5953.9
T4	185.44	4~5 d before 60 d after planting, and 2~3 d from 60 to 120d.	6725.6
T5	231.80	5~7 d before 60 d after planting, and 3~5 d from 60 to 120d.	6344.0

Table 2. Effect of different low irrigation threshold to irrigation resume of Anthurium

As shown in Table 2, the irrigation period expands along with the decreasing of the lower irrigation limit, which is basically same as the conclusion of the routine soil water-saving irrigation test (Yang Wenbin, 2003; Tian Yi et al., 2006); but the irrigation quota of each treatment goes up along with decreasing of the lower irrigation limit. To compare with T1, the quota of T2, T3, T4 and T5 increases 5.3%, 22.7%, 38.2% and 29.3% respectively, which is on the contrary of the routine soil water-saving irrigation test (Sun Huayin et al., 2008). The field water utilization coefficient of the tested substrate goes down along with decreasing of the lower irrigation limit, it is the reason that causes the substrate culture irrigation quota increases along with decreasing of the lower irrigation limit.

3.2.3. *Influence to anthurium growth status from various low irrigation threshold under drip conditions*

Refer to Fig. 5a and Fig. 5b for influences to the plant height and crown diameter of the anthurium from various low irrigation threshold under drip conditions.

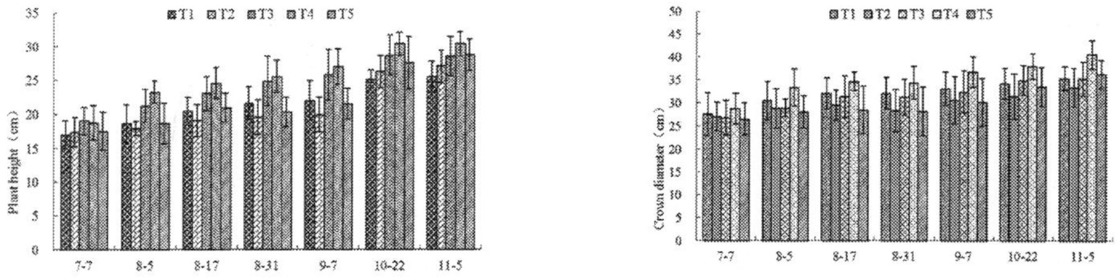

a Plant height b Crown diameter

Figure 5. Effect of different low irrigation threshold to plant height and crown diameter

As Fig. 5a shown, T3 and T4 appear obvious growth vigor. By the end of the test, the plant heights of T1, T2, T3, T4 and T5 are 25.4, 27.0, 28.5, 30.3 and 28.7 cm respectively. Based on the significance testing results, the height of T4 is 19.3% higher than T1 obviously; the differences between other treatments and T1 are not obvious.

As Fig. 5b shown, as with the plant height, the crown diameter of T4 appears obvious growth vigor. By the end of the test, the anthurium crowns of T1, T2, T3, T4 and T5 are 35.3, 33.3, 35.2, 40.5 and 36.2 cm respectively. Based on the significance testing results (p=0.05), the crown diameter of T4 is 14.7% higher than T1 obviously, but the differences between other treatments and T1 are not obvious.

Treatment	Parameter			
	Qty. of spathe	Length of spathe/cm	Width of spathe/cm	Height of inflorescence/cm
T1	2.7±0.8	10.7±1.2	7.5±1.5	3.5±0.4
T2	2.4±0.5	10.4±2.3	7.3±1.5	3.4±1.0
T3	2.5±0.5	10.9±1.5	7.4±1.5	3.7±0.7
T4	2.6±0.4	10.9±1.9	7.8±1.3	4.3±0.7
T5	2.7±0.5	9.8±1.3	7.1±1.0	3.6±1.1

Table 3. Effect of different low irrigation threshold to spathes

Under drip irrigation conditions, refer to Table 3 for influence to the ornamental part - the spathe of the anthurium from various low irrigation threshold. The flower quantity of the 5 treatments are all 3, there is no large difference among treatments. The lengths of the spathe of T1~T4 are all about 10.5cm, the widths are all about 7.5cm, and there is no large difference among treatments (p=0.05). The lowest lower irrigation limit – T5 has the smallest length and width of the spathe, it has negative influence to ornamental quality of the anthurium.

3.2.4. *Influence to anthurium water consumption from various low irrigation threshold under drip irrigation conditions*

Refer to Table 4 for influence to the anthurium water consumption from various low irrigation threshold under drip irrigation conditions.

Treatment	Anthurium water consumption of each treatment/(g·pot^{-1})				
	07-04~07-31	August	September	10-01~11-04	Total water consumption during experiment
T1	572.2	655.5	729.2	1028.9	2985.8
T2	554.4	559.5	599.1	849.3	2562.3
T3	554.4	632.2	696.0	922.4	2805.1
T4	578.8	661.8	699.3	918.4	2858.3
T5	557.3	534.2	538.6	891.7	2521.7

Table 4. Effect of different low irrigation threshold to water consumption amount of Anthurium

Based on Table 4, the water consumption of the tested anthurium during test is between 2 521~2 985 g/pot. To compare with T1, the total anthurium consumption of T2, T3, T4 and T5 are all reduced, in the degree of 14.2%, 6.1%, 4.3% and 15.5% respectively, viz. decreasing of the low irrigation threshold reduce the water consumption of the tested anthurium. Li Xia (Li Xia et al., 2010) carry out research about the transpirations of potted tomato with various substrate moisture contents, they also find that decreasing of the substrate moisture content reduces the transpiration of substrate potted tomato. Difference of water consumption between various treatments is caused by both the plant transpiration and substrate evaporation. Based on the plant heights (Fig. 5 (a)), growth of T1 and T2 is relatively weak; since the substrate moisture content of T1 is always at a higher level because of frequent irrigation, the main reason of large total water consumption is strong substrate evaporation; while the moisture content of T2 is lower than T1 obviously, the main reason of reduced water consumption may be weak substrate evaporation; on account of good growth of T3 and T4, their higher water consumption may be caused by strong transpiration; and the substrate moisture content of T5 is the lowest with weaker growth, substrate evaporation and plant transpiration, which may be the main reason to cause its lower water consumption. So water consumption mechanism of potted substrate needs intensive study.

3.2.5. *Proper lower irrigation limit for soilless culture anthurium under drip irrigation conditions in a multi-span greenhouse*

The characteristics of the flower industry are high input cost and seeking high returns, so the irrigation technology and the irrigation system are crucial to affect flower quality and benefit. Since the economic value of flower depends on quality, the high priority consideration when making the irrigation system shall be high quality of flower. At the same time, multiple benefits such as water saving, fertilizer saving, energy saving and manpower reducing shall be considered synthetically.

In this paper, the lower irrigation limit is key to water consumption and quality of the anthurium. To compare with T1, the flower count and the spathe of T4 that has the lower irrigation limit 60% of the substrate water holding capacity are not affected, but its water consumption has 4.3% reduction, the plant height has 19.3% increasing obviously, the crown diameter has 14.7% increasing obviously, which improves ornamental quality of the anthurium. While to compare T1 with others, though the water consumption goes down to varying degrees, there is no ornamental quality improvement, even negative effect to T5. So under the drip irrigation conditions in this paper, the proper lower irrigation limit of the anthurium shall be 60% of the substrate water holding capacity, here the substrate moisture content is between 60%~80% of the substrate water holding capacity; the irrigation period within 60d of transplanting is 4~5d, for 60~124d it is 2~3d, and the irrigation quota during the test is 185.44g/pot.

Though T4 has better irrigation effect in this paper, its field water utilization coefficient is only about 0.42, with serious leakage loss of good water and fertilizer resource from drip irrigation that become crucial environmental pollution source. It is urgently needed for intensive study on soilless culture substrate water absorption and storage mechanism, research and development of substrate culture drip irrigation technical mode, development of substrate with stable physicochemical property, reinforcement of circulation utilization of good water and fertilizer resource. Improve the utilization efficiency of water and fertilizer, meanwhile guaranteeing flower quality; support the flower industry, meanwhile achieve water-saving irrigation, good product quality and environmental protection and so on.

3.3. Conclusions

(1) The low irrigation threshold is the key controlling factor for the substrate moisture content and the irrigation water use efficiency. Both the substrate moisture content and the field water utilization coefficient go down along with the low irrigation threshold, while the irrigation leakage loss increases along with decreasing of the lower irrigation limit. Under the testing conditions in this paper, the field water utilization coefficient is only about 0.40~0.52, the irrigation period was prolonged along with the low irrigation threshold, but the irrigation quota increases on the contrary. To compare with the treatment that has the prolong as much as 90% of the substrate water holding rate, the irrigation quotas of treatments that have the low irrigation threshold as much as 80%, 70%, 60% and 50% of the substrate water holding rate increase 5.3%, 22.7%, 38.2% and 29.3% respectively.

http://dx.doi.org/10.5772/59466

(2) Under the testing conditions, the difference of the anthurium flower count between various treatments is not large, all of them are 3. When the lower irrigation limit is 60% of the substrate water holding capacity, the spathe size of the anthurium is not affected. But the plant height and the crown diameter increase 19.3% and 14.7% (p=0.05) obviously, in addition to improving ornamental quality. For the other treatments, except for the treatment with the lower irrigation limit as much as 50% of the substrate water holding rate that has reduced anthurium spathe, the others have no obvious difference from the treatment with the lower irrigation limit as much as 90% of the substrate water holding rate.

(3) During the test, the anthurium water consumption is about 2 522~2 986 g/pot, and the water consumption goes down along with decreasing of the lower irrigation limit. To compare with the treatment with the lower irrigation limit as much as 90% of the substrate holding capacity, the water consumptions of the treatments with the lower irrigation limit as much as 80%, 70%, 60% and 50% of the substrate water holding rate reduce 14.2%, 6.1%, 4.3% and 15.5% respectively. In order to achieve the anthurium optimal ornamental quality under the test conditions, determine that the proper lower irrigation limit of the anthurium shall be 60% of the substrate water holding capacity. When the substrate moisture content is between 60%~80% of the substrate water holding capacity, the irrigation period within 60d of transplant is 4~5d, and it is 2~3d after 60d~124d, the irrigation quota is 185.44g/pot.

4. Ebb-and-flow irrigation technology for soilless potted anthurium

4.1. Experimental plant and design

The test starts planting on July 4, 2009 and ends on November 5, takes totally 124d. The pot density to October 14 is 15 pot/m^2, and 10 pot/m^2 after that day. The planted seedlings have uniform growth, about 18cm high. The tested object is potted anthurium, its variety is "Flame", which belongs to *Araceae* and *Anthurium*. The culture substrate is mixture of PINDSTURP peat soil imported from Denmark and perlite from China in a volume ratio of 10:1, its dry volume mass is 0.16 g/cm^3, and the volume water holding rate (θ_{FC}) is 0.4315. For the tested pot, its top diameter is 16cm, its bottom diameter is 10cm, its height is 12.8cm, and its volume is about 1.4L. There are five treatments in the test, from T1 to T4 are ebb-and-flow irrigation, their nutrient solution depths are 1/2, 1/3, 1/4 and 1/5 of the pot height (*H*). T5 is drip irrigation treatment, as shown in Table 5.

Growth stage	Add English version				
	T1	T2	T3	T4	T5
Whole growth period	Nutrient solution depth of $\frac{1}{2}H$	Nutrient solution depth of $\frac{1}{3}H$	Nutrient solution depth of $\frac{1}{4}H$	Nutrient solution depth of $\frac{1}{5}H$	Drip irrigation

Note: *H* is the pot height, here the pot height is 12.8cm.

Table 5. Experimental treatments

The water channel bonding with transparent acrylic sheet is used as the culture vessel of ebb-and-flow irrigation. The tested pots are placed according to the production density. Control the liquid volume according to the set nutrient solution depth for each treatment, after 30min soak, discharge residual water. For drip irrigation treatment, use the pressure compensating dripper with water yield of 3.85 L/h (The working pressure range is 0.05~3 MPa), with the flat perforated pipe configured 4 outlets. Connect the drip arrow with uniform 1m long capillary tube, insert shallowly in the substrate within the center area of the flower root system. Place the pot on the bracket, and put the PVC collector under the bracket to collect water leakage;the lower irrigation limit of each treatment are unified as 80% of the substrate water holding rate (θ_{FC}),but the irrigation quota of drip irrigation takes the upper irrigation limit as 100%θFC. Repeat 5 times per treatment. There are the protective plants around the sample. The self-prime stabilized pressure pump with the pressure tank is used; the hydraulicproportioning fertilization pump made by USA is used for drip irrigation. But for ebb-and-flow irrigation, dissolve fertilizer in water directly for fertilization with water. Determine the times of drip irrigation and fertilization as required by production, the fertilization times of ebb-and-flow is same as drip irrigation. Distribute the 5 test areas randomly, management, protection, pest control, and fertilizer solution concentrations are same.

Test treatment	Irrigation system
T1	The irrigation period for the whole growth period is 5~7 d, with an average of 5.6d. Irrigate nutrient solution till 1/2 of the pot height is reached.
T2	The irrigation period for the whole growth period is 4~6 d, with an average of 4.9d. Irrigate nutrient solution till 1/3 of the pot height is reached.
T3	The irrigation period for the whole growth period is 3~5 d, with an average of 3.7d. Irrigate nutrient solution till 1/4 of the pot height is reached.
T4	The irrigation period for the whole growth period is 3~5 d, with an average of 3.7d. Irrigate nutrient solution till 1/5 of the pot height is reached.
T5	The irrigation period for the whole growth period is 2~3 d, with an average of 2.6d. And the irrigation quota is $92.72 g \cdot pot^{-1}$.

Table 6. Irrigation system for all treatments

4.2. Results and discussion

4.2.1. Dynamics of substrate moisture content

Refer to Fig. 6 for substrate moisture content after irrigation of each treatment. For all treatments, the substrate moisture content after irrigation appears slight fluctuation. The ranges of substrate moisture content after irrigation for T1, T2, T3, T4 and T5 are 106.5%~118.2%θ_{FC}, 96.7%~110.0%θ_{FC}, 90.2%~103.4%θ_{FC}, 89.4%~105.0%θ_{FC} and 82.7%~93.1%θ_{FC} respectively. The mean values of substrate moisture contents after irrigation are 112.8%θ_{FC}, 105.8%θ_{FC}, 95.9%θ_{FC}, 96.8%θ_{FC} and 89.1%θ_{FC} respectively. Based on those, the average relative substrate moisture content after ebb-and-flow irrigation is 6.8~23.7%θ_{FC} higher than that of drip irrigation. The moisture contents of T1 and T2 are higher than the other 3 treatments obviously. T1 is the highest, while T5 with drip irrigation treatment is the lowest; the moisture contents of T3 and T4 with ebb-and-flow irrigation are not very different. When the nutrient solution depth does not exceed 1/4 of the pot height, the influence from the nutrient solution depth to the relative substrate moisture rate after irrigation is not large; but when it exceeds 1/4 of the pot height, the influence increases obviously along with increasing of the nutrient solution depth.

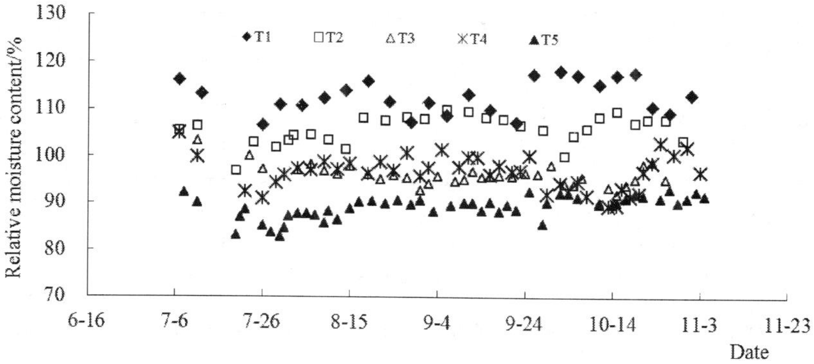

Figure 6. Moisture content after irrigation for all treatments

Refer to Table 7 for the substrate moisture contents analysis during the test. The average relative substrate moisture contents of T1, T2, T3, T4 and T5 are 98.5%θ_{FC}, 95.4%θ_{FC}, 90.4%θ_{FC}, 90.7%θ_{FC} and 84.7%θ_{FC} respectively. The ebb-and-flow is 5.7%~13.8%θ_{FC} higher than drip irrigation. Both of them have the phenomenon that the moisture content exceeds the substrate water holding rate. The moisture content of drip treatment T5 is lower than its substrate water holding rate. So 1/4 of the pot height is the critical value at which nutrient solution may affect the substrate moisture content. When the nutrient solution depth does not exceed 1/4 of the pot height, there is no large influence to the substrate moisture content, but when it exceed 1/4 of the pot height, the proportion of number of days when the moisture content exceeds 100%θ_{FC} increases along with increasing of the nutrient solution depth, while the number of days when it is between 80%~100%θ_{FC} is relatively decreasing.

The substrate moisture content is the main factor of influence for the substrate water, air and temperature environment. Because organic substrate is different from soil, absorption, research on storage and migration of water in substrate is not intensive yet. Qi Haiying (Qi Haiying et al., 2009) find that the substrate pore structure and volume are obviously affected by various water supply modes. Furthermore, pore structure transformation may cause change of water holding capacity. It may be drip arrow local water supply affecting the tested culture substrate volume and pore structure that cause the relative substrate moisture content is lower than the substrate water holding rate after drip irrigation (Ma Fusheng et al., 2012). It is necessary to study further for the law of the pore structure change for soilless culture substrate and its influence to the substrate water absorption, storage and dissipation under various water supply conditions.

Moisture content	Item	Treatment				
		T1	T2	T3	T4	T5
>100%θ_{FC}	Number of days/d	54	43	7	6	0
	Percentage in total test days/%	43.5	34.6	5.6	4.7	0
80%~100%θ_{FC}	Number of days/d	70	81	117	118	124
	Percentage in total test days/%	56.5	65.4	94.4	95.3	100
Average substrate moisture content during the test/(%θ_{FC})		98.5	95.4	90.4	90.7	84.7

Table 7. Level of substrate moisture content for all treatments

4.2.2. Irrigaiton timing and irrigation water use efficiency

In this paper, the irrigation periods of all treatments during the test are 5~7 d (Average 5.6 d) for T1, 4~6 d (Average 4.9 d) for T2, 3~5 d (Average 3.7 d) for T3 and T4, and 2~3 d (Average 2.6 d) for T5 respectively. Based on those, when all the low irrigation threshold are 80% of the substrate water holding rate, the irrigation period of the ebb-and-flow irrigation is longer than drip irrigation. The average irrigation periods of T1, T2, T3 and T4 are 3d, 2.3d and 1d longer than T5, the treatment of drip irrigation, which is in conformity with the law of substrate moisture content change. So 1/4 of the pot height is the critical value of the nutrient solution depth at which the ebb-and-flow irrigation period may be affected. When the nutrient solution depth does not exceed 1/4 of the pot height, there is no large difference among irrigation periods, but when it exceed 1/4 of the pot height, the irrigation period is prolonged along with increasing of the nutrient solution depth. Viz. under the conditions of controlling the same low irrigation threshold, ebb-and-flow irrigation may prolong the irrigation period and reduce manpower. Catherline A. Neal (Catherline A. Neal et al, 1993) find the irrigation period of ebb-and-flow irrigation is longer than drip irrigation, the same result as this paper.

As the main production form of facility flower, substrate potting mainly adopt walking tube-well irrigation, drip irrigation and other traditional water supply technology. The drip irrigation field water utilization coefficient of tested substrate in this paper is only 0.4~0.5 (Ma Fusheng et al., 2012), which is even lower than the limit of 0.9 specified in *Technical code for microirrigation engineering* (GB/T 50485-2009), there is serious clean and good water and fertilizer leakage loss. This may be affected by the dripper flow. According to Li Mingsi and others' (Li Mingsi et al., 2006) study, there is obvious influence to the soil wetting pattern. But the impact of law from the dripper flow on the organic culture substrate wetting pattern with finite volume needs to be study further. With the help of the ebb-and-flow irrigation technology, equipment such as residual water recycling, water treatment and recycling can be used to achieve circulation utilization of water and fertilizer resources, so the water and fertilizer utilization efficiency can be more than 90% (Zhang Xiaowen et al., 2011). To compare with drip irrigation, ebb-and-flow irrigation may improve water and fertilizer utilization efficiency, and reduce their leakage loss significantly.

4.2.3. Water consumption of anthurium

Refer to Fig. 7 for changing process of water consumption rate of anthurium in the 5 treatments. The water consumption rate of the tested anthurium increases gradually with fluctuation along with extension of growth and development. During the test, the Max. water consumption rates of T1, T2, T3, T4 and T4 are 52.3, 47.8, 46.8, 45.2 and 33.6 g/ (pot d) respectively, while the Min. water consumption rates are 12.3, 11.2, 11.3, 9.0 and 7.0 g pot^{-1} d^{-1} respectively, and the averages are 35.1, 31.7, 30.6, 29.8 and 21.1 g pot^{-1} d^{-1}. All the water consumption rates of ebb-and-flow irrigation treatment of anthurium are higher than drip irrigation.

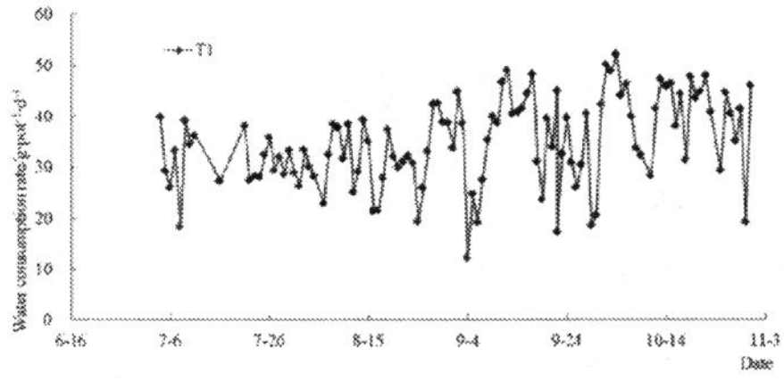

a. T1(Ebb-and-flow irrigation, the nutrient solution depth is 1/2 of the pot height)

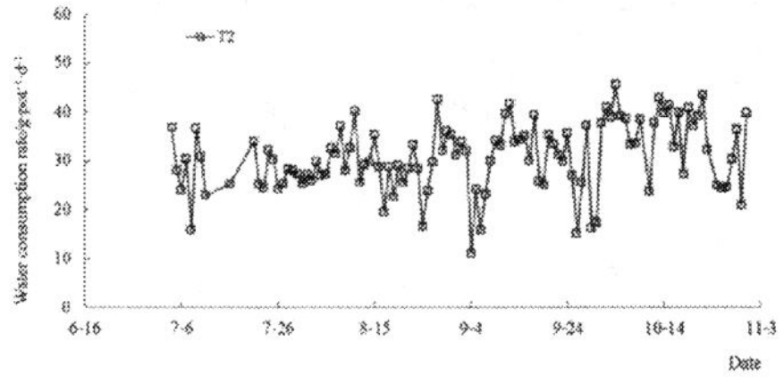

b. T2 (Ebb-and-flow irrigation, the nutrient solution depth is 1/3 of the pot height)

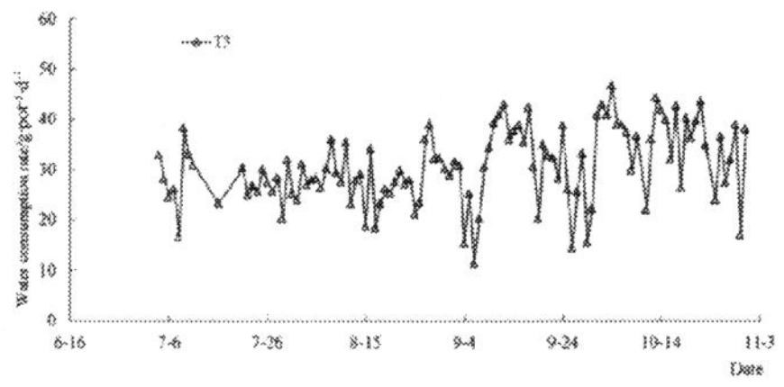

c. T3 (Ebb-and-flow irrigation, the nutrient solution depth is 1/4 of the pot height)

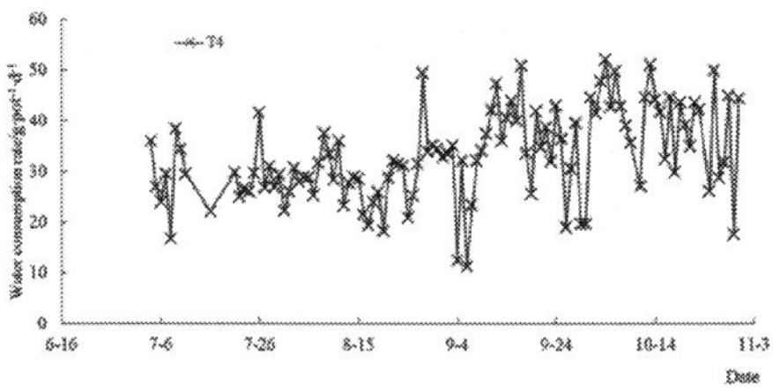

d. T4 (Ebb-and-flow irrigation, the nutrient solution depth is 1/5 of the pot height)

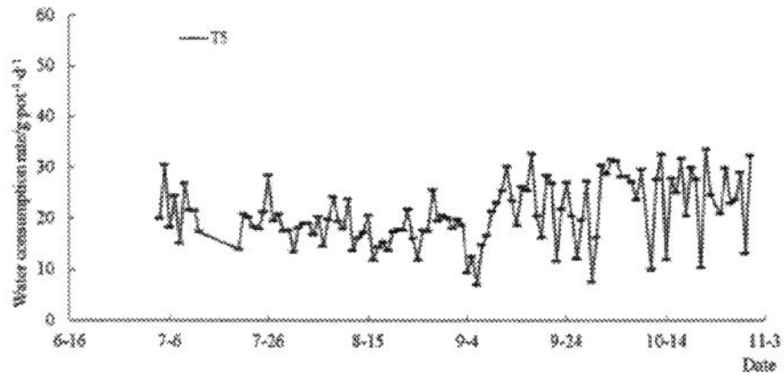

e. T5 (Drip irrigation)

Note: Count by pot.

Figure 7. Daily water consumption for all treatments during experiment

Refer to Table 8 for the monthly average water consumption rate and water consumption during the test. From July to October, the water consumption rate and water consumption of each treatment increase along with plant growth. The irrigation modes affect water consumption of anthurium obviously. The average water consumption rate, monthly water consumption and total water consumption of ebb-and-flow irrigation are higher than drip irrigation obviously during the test, the total water consumptions of T1, T2, T3 and T4 increase 69.3%, 53.2%, 47.5% and 44.2% than T5, so the variation trend of increasing along with increasing of the nutrient solution depth generally.If the nutrient solution utilization efficiency in ebb-and-flow irrigation is 0.9, and in drip irrigation is 0.5, nutrient solution used in ebb-and-flow irrigation during the test is 4 106.6~4 820 g pot^{-1} d^{-1}, 6%~20% lower than drip irrigation (5 124.6 g pot^{-1} d^{-1}), it is reduced distinctly.

Item	Treatment	Time				
		07-04-07-31	08-01-08-31	09-01-09-30	10-01-11-04	Total
Average water consumption rate /($g{\cdot}pot^{-1}{\cdot}d^{-1}$)	T1	30.8	32.0	34.2	41.7	-
	T2	27.6	29.6	30.5	37.8	-
	T3	26.7	28.3	29.9	36.0	-
	T4	27.1	28.5	26.7	35.8	-
	T5	18.1	18.0	20.0	25.7	-

Item	Treatment	Time				
		07-04-07-31	08-01-08-31	09-01-09-30	10-01-11-04	Total
Water consumption/ ($g \cdot pot^{-1}$)	T1	861.8	991.4	1026.3	1458.2	4337.6
	T2	771.8	918.9	914.1	1321.8	3926.7
	T3	747.4	876.6	896.8	1259.5	3780.2
	T4	758.6	884.9	801.0	1251.4	3695.9
	T5	505.6	559.5	599.1	898.0	2562.3

Table 8. Daily and total water consumption amount of Anthurium for all treatments from 07-04 to 11-04

4.2.4. Growth and quality of anthurium

The plant height and the crown diameter are the key indexes concerned ornamental quality of the anthurium, refer to Fig. 8 for the plant height and the crown diameter of the anthurium in each treatment. By the end of the test, the plant heights of T1, T2, T3, T4 and T5 are 33.6, 34.7, 35.3, 38.4 and 27.0 cm respectively, the crown diameters are 41.9, 42.6, 40.6, 43.9 and 33.3 cm respectively. The plant heights and the crown diameters from ebb-and-flow treatment are better than drip irrigation treatment obviously (p=0.05). The plant height of T4 in ebb-and-flow treatment is 14.3% higher than T1 (p=0.05), but the plant heights among other treatments and the crown diameters of all treatments have no obvious difference.

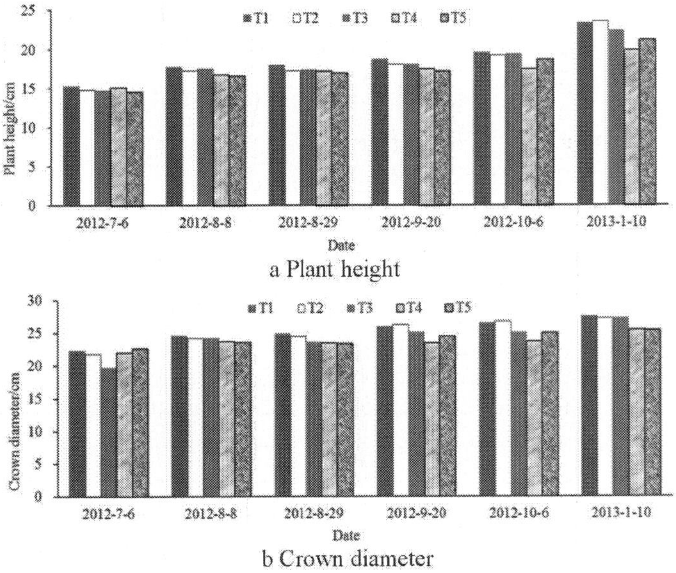

Figure 8. Plant height and crown diameter for all treatments

Refer to Table 9 for the spathe of each treatment. The spathe is the main ornamental part of the anthurium. By the end of the test, the quantity difference of the spathe in each treatment is not large, all are about 3. Compared with T5 of drip irrigation, the length of the spathe with ebb-and-flow irrigation increases 0.8~1.5 cm, the width of it increases 0.5~1.8 cm, except for the inflorescence height of T4 with ebb-and-flow irrigation treatment is about 1cm higher than T5, the inflorescence heights of other treatments have no large difference from T5. There is no obvious difference among treatments (p=0.05).

Treatment	Parameters			
	Qty. of spathe/pc	Length of spathe/cm	Width of spathe/cm	Height of inflorescence/cm
T1	3.0	11.2	7.8	3.7
T2	2.6	11.5	8.1	3.6
T3	2.4	11.9	8.2	3.7
T4	3.4	12.9	9.1	4.4
T5	2.4	10.4	7.3	3.4

Table 9. Spathes of all treatments

In conclusion, when the low irrigation threshold are controlled to 80% of the substrate water holding rate, growth of the anthurium is affected obviously from irrigation modes. Anthurium growth of ebb-and-flow treatment is better than that from drip irrigation treatment. The anthurium plant height and spathe size of T4 with the nutrient solution depth being 1/5 of the pot height appear uniform superiority.Catherine A. Neal (Catherine A. Neal et al. 1992) find plant growth is affected by irrigation modes, while ebb-and-flow irrigation is a technology that may improve irrigation water utilization efficiency and get potential optimal growth. Interaction between water and fertilizer is an important factor for plant growth. John M. Dole (John M. Dole et al, 1994) find the Poinsettias with ebb-and-flow irrigation has the best water utilization efficiency, plant height, stem diameter, leaf width and total amount of dry matter, in addition, its growth is affected by the nutrient concentration of the nutrient solution. Daniel I. Leskovar (Daniel I. Leskovar et al, 1998) find the irrigation mode, water regime and fertilizer application are key for growth of the plant root system and sprouting, but their interaction mechanism needs intensive study. Ma Fusheng (Ma Fusheng et al., 2011) find the lower irrigation limit affect anthurium growth even under drip irrigation conditions. Influence to anthurium growth from various irrigation modes in this paper may be caused by water and fertilizer coupling mechanism of each treatment, the influence mechanism from soilless culture water and fertilizer utilization to plant growth needs intensive study.

4.2.5. Suitable irrigation system for anthurium

According to experimental results, the quantity, length and width of the spathe, in addition to the inflorescence height of T4 with the nutrient solution depth as much as 1/5 of the pot height are the best. In addition, its water consumption is 15% less than T1. So it is the suitable nutrient solution control depth with ebb-and-flow irrigation, here the nutrient solution depth is 2.56cm, identify with the result of 25mm for most flower nutrient solution depth recommended by Yang Tieshun (Yang Tieshun et al., 2009). So in the testing conditions of this paper, in order to get optimal anthurium quality, the 1/5 pot height shall be taken as the ebb-and-flow irrigation nutrient solution depth, viz. 2.56cm, the irrigation period is affected by water, fertilizer, air and temperature environment of flower, it is 3~5d.

4.3. Conclusions

(1) The irrigation technologies are key factor to the substrate moisture content, the irrigation period and the water consumption law. The substrate moisture content after ebb-and-flow irrigation is between 96%~113%θ_{FC}, increase 7%~23%θ_{FC} than 89.1%θ_{FC} of drip irrigation; the average substrate moisture content during the ebb-and-flow test is 90.4%~98.5%θ_{FC}, increase 5.7%~13.8%θ_{FC} than 84.7%θ_{FC} of drip irrigation treatment. The anthurium water consumption with ebb-and-flow irrigation is 3 696~4 338 g pot^{-1} d^{-1}, increase 44.2%~69.3% than 2 562.3 g pot^{-1} d^{-1} of drip irrigation; the irrigation water utilization increases from 0.4~0.5 of drip irrigation to 0.9, the irrigation nutrient solution reduces from 5124.6g/pot to 4106.6~4820.0g/pot; the irrigation period of ebb-and-flow irrigation is 3~7 d, extending 1~3 d than drip irrigation in average, which improve the labor productivity.

(2) The irrigation modes have significant influence to quality of the anthurium. The plant height of the anthurium in ebb-and-flow irrigation treatment is 33.4~38.6 cm, 6.6~11.4 cm higher than 27.0 cm in drip irrigation, the crown diameter is 40.6~43.9 cm, 7.3~10.6 cm higher than 33.3 cm in drip irrigation; the length and width of the spathe are 11.2~12.9~cm and 7.8~9.1 cm, 0.8~1.5 cm and 0.5~1.8 cm higher than 10.4cm (Length) and 7.3cm (Width) in drip irrigation respectively, the inflorescence height of the spathe is 3.6~4.4 cm, 0.2~1.0 cm higher than 3.4cm in drip irrigation treatment.

(3) The 1/4 pot height is the critical value of the nutrient solution depth that affects the substrate moisture content, the anthurium water consumption and the irrigation period of ebb-and-flow irrigation. When the nutrient solution depth is not more than the 1/4 pot height, there are no large differences among the substrate moisture content, the anthurium water consumption and the irrigation period; but when it is more than the 1/4 pot height, the substrate moisture content, the anthurium water consumption and the irrigation period soar along with increasing of the nutrient solution depth.

(4) When the ebb-and-flow irrigation nutrient solution depth reaches the 1/5 pot height, the ornamental value of the anthurium is optimal, furthermore, the water consumption reduces 15% than that of T1 at which the nutrient solution depth is the 1/2 pot height. The total irrigation water consumption reduces 20% than drip irrigation. Here the nutrient solution depth is 2.56cm, the irrigation period is between 3~5d, the average value is 3.7d. So 2.56cm can be the better nutrient solution depth under the testing conditions.

5. Experimental research on suitable dripper discharge for potted anthurium with soilless culture

5.1. Experimental plant and design

The anthurium is chosen as the testing object; its variety is "Alabama", which belongs to *Araceae* and *Anthurium*. The culture substrate is PINDSTURP peat soil imported from Denmark, with the dry volume mass of 0.16g/cm^3, and the water holding rate (θ_{FC}) of 0.4315 (V/V). The tested anthurium is cultivated in plastic pot with 16cm of top diameter and 12.8cm of height, the volume is about 1.4L, and the substrate is filled to the pot evenly. Refer to Table 10 for testing treatment.

Testing stage			Dripper flow/(L/h)		
T1	T2	T3	T4	T5	
Whole growing period	0.55	0.95	1.1	2.2	3.8

Table 10. Experimentle treatments

Note: T1 adopts the plain end water separator configured 4 outlets with the flow of 2.2L/h, T2 adopts the plain end water separator configured 4 outlets with the flow of 3.8L/h, T3 adopts the plain end water separator configured 2 outlets with the flow of 2.2L/h, T4 adopts the 2.2L/h dripper, and T5 adopts the 3.8L/h dripper, the capillary tube after the distributive pipe connects the drip arrow.

In this paper, all treatment irrigation quota are controlled by the substrate water holding capacity as the upper irrigation limit, and 60% of the substrate water holding capacity as the lower irrigation limit. The test starts on June 6, 2012 and ends on January 12, 2013, takes 221d totally. The placing density of tested anthurium plantlet is 27 pot/m^2; and 23 pot/m^2 from August 9, 2012 to the end of the test. The plantlet is 16cm in height with uniform growth when transplanting. For all treatments, connect the drip arrow with uniform 1m long capillary tube, insert shallowly in the pot substrate from the side direction about 1cm within the center area of the root system. Place the pot on the bracket. Repeat 7 times per treatment. There are the protection lines at both sides of the monitored sample plant. The self-prime stabilized pressure pump with the pressure tank is used for driving; the hydraulicproportioning fertilization pump made by USA is used for fertilization with water.All test plots are distributed randomly, the treatment measures such as farming, fertilizing, and pest control are same.

5.2. Results and discussion

5.2.1. Influence to water consumption of soilless potted anthurium from emitters with various dripper flows

The water consumption rate of tested anthurium was analyzed in this paper. Table 11 shows the water consumption rate of each treatment is between 8~24g pot^{-1} d^{-1} in June and July, and between9~32 g pot^{-1} d^{-1} in August and September, reaching the peak of water consumption 12~36g pot^{-1} d^{-1} in October, and falling back to 8~24g pot^{-1} d^{-1} in November and December. The change process from June to December is rise-fall. Refer to Table 11 for the average monthly water consumption rate and the water consumption per treatment.

Item	Treatment	Month							
		6	7	8	9	10	11	12	Total
Average water consumption rate/ $g·pot^{-1}·d^{-1}$	T1	16.6	17.3	20.5	20.0	22.6	17.3	17.1	——
	T2	16.0	16.9	19.4	19.0	21.9	16.8	17.3	——
	T3	16.0	17.0	19.7	18.9	22.7	17.0	17.3	——
	T4	15.9	15.7	17.7	17.4	21.5	16.4	16.7	——
	T5	15.8	16.1	18.9	17.7	21.0	16.6	16.7	——
Water consumption/ $g·pot^{-1}$	T1	414	479	637	599	679	485	529	3823
	T2	400	466	609	571	657	471	536	3709
	T3	400	472	612	567	680	476	535	3743
	T4	397	437	548	522	643	459	517	3523
	T5	396	448	585	530	629	464	517	3569

Table 11. Daily and total water consumption amount of all treatments

According to Table 11, the water consumptions of each treatment in the whole growth period are 3823g pot^{-1} of T1, 3709g pot^{-1} of T2, 3743g pot^{-1} of T3, 3523g pot^{-1} of T4 and 3569g pot^{-1} of T5, the variation trend goes down along with increasing of the dripper flow. The water consumption of T1 with the Min. dripper flow is the highest, which is about 8% more than T5 (The Max. dripper flow) and T4 (The second largest dripper flow). This may be that the dripper with a small flow can establish a higher substrate moisture content, thereby creates more favorable transpiration environment. The water consumptions of T2 and T3 (Their dripper flows are not different greatly) are also close. Though the dripper flow of T4 is far lower than T5, there is no large difference between their water consumption. So increasing the anthurium water consumption by reducing the dripper flow has obvious effect only in a certain range and certain gradient. Wang Xiukang (Wang Xiukang et al., 2010) find the influence to corn root system from the dripper flow. When 1L h^{-1}~2.5L h^{-1}is chosen as the dripper flow, the dripper flow of 2L h^{-1}has obvious influence to spatial distribution of corn root system.

5.2.2. *Influence to irrigation period of soilless potted anthurium from emitters with various dripper flows*

Refer to Table 12 for influence of soilless potted anthurium irrigation period (*T*) from emitters with various dripper flows. The irrigation period goes up along with decreasing of the dripper flow, so the manpower cost can be saved.

Treatment	T1	T2	T3	T4	T5
T (d)	8.3	8.1	8.1	7.6	6.3

Table 12. Effect of different dripper discharge to water use efficiency and irrigation cycle of Anthurium

5.2.3. Influence to anthurium growth and quality from emitters with various dripper flows

There is no obvious difference on the tested anthurium spathe by the end of the test. Analyze the monitoring information of July 6, August 8, August 29, September 20, October 6 and December 31. Refer to Fig. 9 (a) for influence to tested anthurium plant height from various dripper flows, and to the crown diameter refer to Fig. 9 (b).

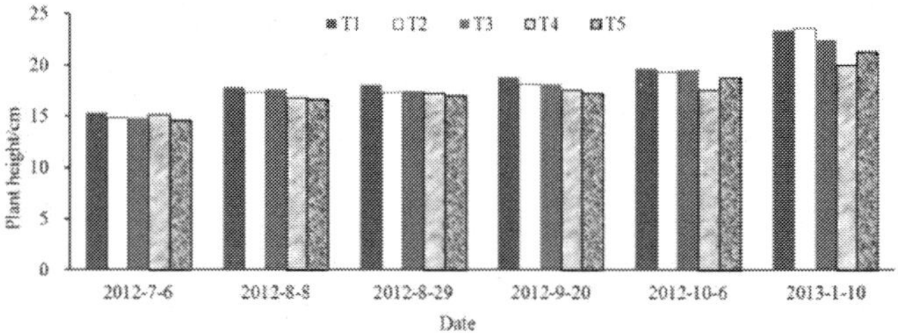

a Plant height

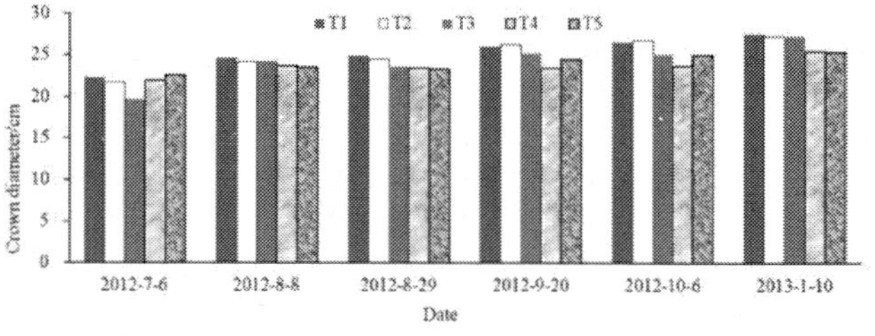

b Crown diameter

Figure 9. Effect of different low irrigation threshold to plant height and crown diameter

Based on Fig. 9, there is no large difference on the anthurium plant height and the crown diameter under the treatment conditions of T1, T2 and T3 (With lower dripper flows), all of them have obvious growth vigor. The plant heights and the crown diameters of T4 and T5 are lower than the first 3 treatments. By the end of the test, the plant heights of each treatment are

http://dx.doi.org/10.5772/59466

23.3cm of T1, 23.5cm of T2, 22.4cm of T3, 19.9cm of T4 and 21.2cm of T5; the crown diameters are 27.5cm of T1, 27.2cm of T2, 27.3cm of T3, 25.4cm of T4 and 25.3cm of T5.

5.2.4. *Selection of drip irrigation emitters suitable for tested soilless potted anthurium*

Based on comprehensive analysis on influence to the tested anthurium substrate moisture content, anthurium water consumption, field water utilization coefficient, irrigation period and anthurium quality from various dripper flows, to compare with T5, the anthurium water consumption of T1~T3 increases 5%~8%, and the irrigation period prolongs 1~2d, in addition to better ornamental effect on the anthurium plant height and crown diameter, especially the comprehensive advantage of T1. So within the range of the tested dripper flow, the dripper flow of 0.55L h^{-1} is recommended for tested soilless potted anthurium, with the irrigation period of 8.3d. Because of the limited research conditions in this paper, the suitability to tested anthurium irrigation if the dripper flow continues to reduce needs further investigation.

The low irrigation threshold in this paper is 60%θ_{FC}. Ma Fusheng (Ma Fusheng et al., 2012) and others find there is obvious law of influence to the field water utilization coefficient of soilless potted anthurium under the same dripper flow in various low irrigation threshold, the higher lower irrigation limit, the larger field water utilization coefficient. Selection and supporting technology of dripper flow under various low irrigation threshold needs further study. Good water and fertilizer resource leakage loss in drip irrigation is serious, and becomes major environmental pollution source, so it is urgent to research the water absorption and storage mechanism of soilless culture substrate, develop the substrate culture drip irrigation technical mode, develop the substrate materials with stable physicochemical properties, improve cyclic utilization of good water and fertilizer resource, improve the utilization efficiency of water and fertilizer resource meanwhile guarantee flower quality, in order to support the flower industry for the achievement of water-saving irrigation, good product quality, industrial environment protection and so on.

5.3. Conclusions

(1) The water consumption rate of the tested anthurium increases from 8~24g pot^{-1} d^{-1} in June to 12~36g pot^{-1} d^{-1} in October, and then fall back to 8~24g pot^{-1} d^{-1} in December; the water consumption of the whole growth period per treatment is 3569g pot^{-1}~3823g pot^{-1}, with the change trend of decreasing along with the increasing of the dripper flow.

(2) Under the 3 low-flow drippers with the flows of 0.55L h^{-1}, 0.95L h^{-1} and 1.1L h^{-1}, the plant height and the crown diameter of the anthurium have no large difference, but all of them are higher than the two treatments of 2.2L h^{-1} and 3.85L h^{-1}, which affects the spathe largely. So a dripper with the small flow is conductive to good quality.

(3) The small dripper flow may prolong the irrigation period for about 2d, and get better ornamental quality. To compare and select the tested drippers based on the economic benefit and water utilization efficiency, the small-flow dripper is suitable for the soilless substrate potted flower.

6. Research on moisture characteristic parameters for soilless culture substrate

6.1. Test design

Perform the single factor test with typical soilless culture substrates, and use sandy soil and sandy loam soil for the contrast test. There are 10 treatments, domestic peat (T1), vermiculite (T2), perlite (T3), peat imported from Germany (T4), domestic peat and vermiculite in 2:1 of mass ratio (T5), domestic peat and perlite in 2:1 of mass ratio (T6), domestic peat, vermiculite and perlite in 1:1:1 of mass ratio (T7), domestic peat, vermiculite and perlite in 3:1:1 of mass ratio (T8), sandy soil (CK1), sandy loam soil (CK2). Because perlite and peat imported from Germany have large particle, so the air-dried volumetric specific gravities when filling are 0.12 g/cm^3 and 0.15 g/cm^3 respectively, sandy soil and sandy loam soil are 1.4g/cm^3, the other treatments are 0.25 g/cm^3. Determine the moisture characteristic parameters such as the porosity, the permeability coefficient, the water characteristic curve, the diffusivity and so on for each treatment. Repeat 3 times per treatment.

6.2. Results and analysis

6.2.1. Porosity

During practical production of seedling raised with substrate, the total porosity is normally 70%~90%, the water-air ratio is normally 2.0~4.0, which may meet demands of crop to moisture and oxygen (Li Douzheng, 2006). But a single substrate is hard to meet all those requirements at the same time. Refer to Table 13, the water holding porosity of vermiculite (T2) is large, viz. vermiculite has good water holding capacity, but poor aeration porosity, so its water-air ratio is unreasonable. The perlite (T3) has large aeration porosity, viz. perlite has good air permeability, but poor water holding capacity. So in the mixtures of T5, T6, T7 and T8, vermiculite is used to improve its water holding capacity, and perlite is used to improve the air permeability.

	T1	T2	T3	T4	T5	T6	T7	T8	CK1	CK2
Total porosity/%	80.11	81.10	82.41	81.82	78.09	82.45	78.42	84.73	50.74	40.55
Aeration porosity/%	18.38	14.91	37.10	18.01	19.51	21.90	20.70	21.90	2.84	1.94
Water holding porosity/%	61.73	66.19	45.31	63.81	58.58	60.55	57.72	62.83	47.90	38.61
Water-air ratio	3.35	4.44	1.22	3.54	3.00	2.77	2.79	2.87	16.87	19.90

Table 13. Comparison of the different height containers

Refer to table 13, the porosity of the substrate is higher than sandy soil (CK1) and sandy loam soil (CK2) obviously, this is due to the particle of sandy soil and sandy loam soil is small, and the unit weight is considerably higher than that of the substrate. Thereby the porosity of sandy soil and sandy loam soil is small. To compare with sandy soil and sandy loam soil, moisture in substrate is easier to be used by plant, to compare the above mentioned 4 mixed substrate, T8 has the Max. porosity of 84.7%, the water-air ratio of 2.9, so it has large total porosity and good air permeability, and a reasonable water-air ratio.

6.2.2. Analysis of permeability coefficient

The permeability coefficient is an important index to affect plant growth, a permeability coefficient suitable for plant growth depends on the soil type. Normally sandy loam soil, loam and clay are suitable for corn growth. The permeability coefficients of the 3 kinds of soil are 6×10^{-2}, 6×10^{-3} and 6×10^{-4}mm min^{-1}(Du Yanling et al., 1992). Currently, there is seldom research on the permeability coefficient of a substrate. If the permeability coefficient is too high, the substrate is difficult to hold water to cause leakage after irrigation and waste water resource, if the permeability coefficient is too low, the air permeability of the substrate is bad, which may affect breath and growth of the root system (Shi Lianhui et al., 2008).

Refer to Fig. 10, except for T3 and CK1, the standard deviations of all treatments are all 0.2%~10% without large fluctuation. It means this determination method for the permeability coefficient can represent the permeability coefficient of a substrate. The permeability coefficient of sandy loam soil is 5×10^{-2} mm/min, which can meet requirement for corn growth. After comparing the permeability coefficients of CK2 and the substrate, the permeability coefficient of the substrate is higher than CK2 obviously for the aeration porosity of the substrate is larger, and water infiltration rate is faster than sandy loam soil after irrigation. So it may reach the corn water demand quickly, and little and frequent irrigation is appropriate. By comparing the permeability coefficients of substrates, the permeability coefficient of T2 is the largest, 2.20 mm min^{-1}, while T6 is the smallest. In addition, the permeability coefficient of a single substrate is higher than the composite substrate obviously, for there are more small particles in the composite substrate to blocking the pores in the substrate to slow down the permeation. So if a single substrate is used for culture and seedling, the irrigation frequency shall be adjusted reasonably.

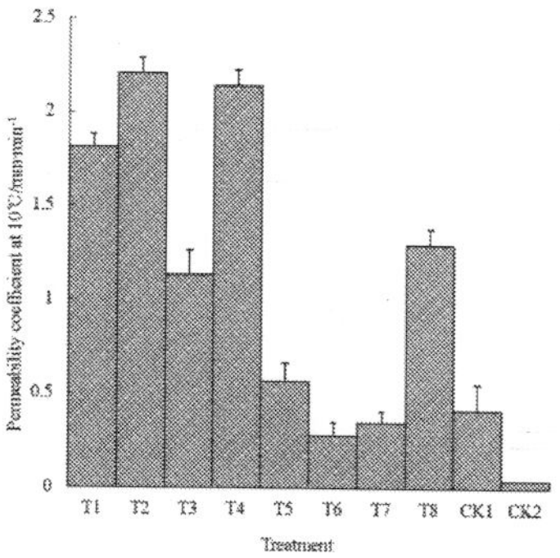

Figure 10. Comparison of permeability coefficient

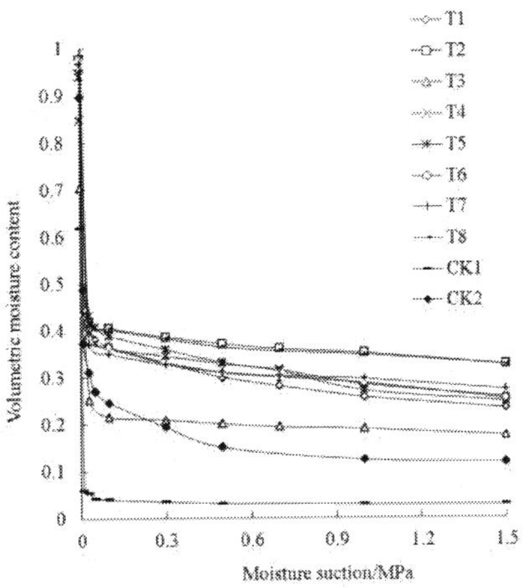

Figure 11. Comparison of moisture retention curves

In the drip irrigation engineering design, the dripper spacing is an important index to affect irrigation evenness, while the permeability coefficient is an important index to affect the dripper spacing (Yao Zhenxian et al., 2012). The dripper spacing suitable for the substrate shall be determined according to the dripper flow in order to improve the water utilization efficiency of the drip irrigation system.

6.2.3. *Water characteristic curve*

As shown in Fig. 11, at the lower suction stage at which the moisture suction is less than 0.1Mpa, the volumetric moisture contents of 10 treatments reduced sharply along with increasing of suction. The water holding capacity of T8 and T2 is the largest, and CK1 is the smallest. At the stage with the moisture suction is more than 0.1Mpa, the volumetric moisture contents reduce a little along with increasing of suction. Refer to Fig. 4 for the moisture content distribution, within the effective water range, viz. the substrate moisture suction is 0.01-0.1Mpa, the standard deviation of each treatment is 0.8%~3% with small fluctuation, it means the water characteristic curve can represent the water characteristic curve of the substrate.

In recent years, Kang Yuehu and others control the irrigation water suitable for growth and moisture efficient utilization of vegetables such as Chinese cabbage, cowpea and tomato under drip irrigation by controlling the soil matric potential. They guide irrigation of vegetableswithin the lower soil matric potential limit of -0.02~-0.05Mpa (Jiang Shufang et al., 2009; Zhang Chao et al., 2010; Wan Shuqin et al., 2009; Theodore W Sammis et al, 1980; Jia Junshu et al., 2011; Riviere L M et al, 2001). At the same time, take the volumetric moisture content within the ranges of 0.01~0.1Mpa, 0.01~0.05Mpa and 0.05~0.1Mpa of substrate water suction as the basis for division of substrate available water, easy available water and buffer water.

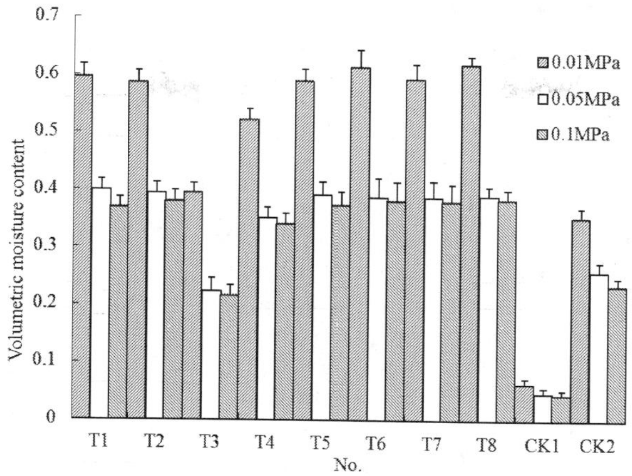

Figure 12. Comparison of water content between 0.01and 0.1 Mpa

The available water, easy available water and buffer capacity are important indexes to judge water available for absorption by plant and guide the irrigation frequency (Shi Lianhui et al., 2008). The available water, easy available water and buffer capacity of 8 substrates may be obtained by the known water characteristic curve as Table 14 shown.

	T1	**T2**	**T3**	**T4**	**T5**
Available water	22.7	20.8	17.9	18.1	21.7
Easy available water	19.6	19.3	17.1	17.1	19.8
Buffer water	3.1	1.5	0.8	1.0	1.9
	T6	T7	T8	CK1	CK2
Available water	23.3	21.3	23.5	2.0	11.7
Easy available water	22.6	20.6	22.9	1.7	9.3
Buffer water	0.7	0.7	0.6	0.3	2.4

Table 14. Available water, easy available water and buffer capacity

As shown in Table 14, the available water capacity of substrate is higher than that of CK1 and CK2 obviously, which means the available water in the substrate is more higher, so crop may absorb more water in the substrate than in sandy soil or sandy loam soil. To compare the 8 substrates, both available water and easy available water in T8 are the highest, secondly is T6, which means the available water content in T8 and T6 is higher, so crop may absorb more available water. Their irrigation frequencies can be reduced than other substrates.

6.2.4. *Analysis of horizontal diffusivity*

As Fig. 13 shown, when θ<0.3, the diffusivity changes slowly along with increasing of the moisture content, moisture in the substrate is mainly vapor movement [31] (Fan Yanwei, 2008); when θ>0.3, the diffusivity soars along with increasing of the moisture content, a higher substrate moisture content is in favor of substrate diffusive motion; when θ=0.3, refer to Table 3 for the diffusivity of each treatment.

When θ=0.3, the diffusivity of sandy soil is the largest, 654.75 cm^2 min^{-1}, which means the horizontal suction and seepage velocity is quick, moisture spreads horizontally quickly; the second is perlite, 249.15 cm^2 min^{-1}. Because perlite is lighter with smaller unit weight and larger particle size, moisture spreads quickly in perlite; to compare with sandy loam soil, the diffusivity of the substrate is higher, which means the horizontal diffusion velocity in substrate is quick. Thereby when using substrate as breeding and culture substrate, drip irrigation may improve the evenness and be in favor of crop water absorption.

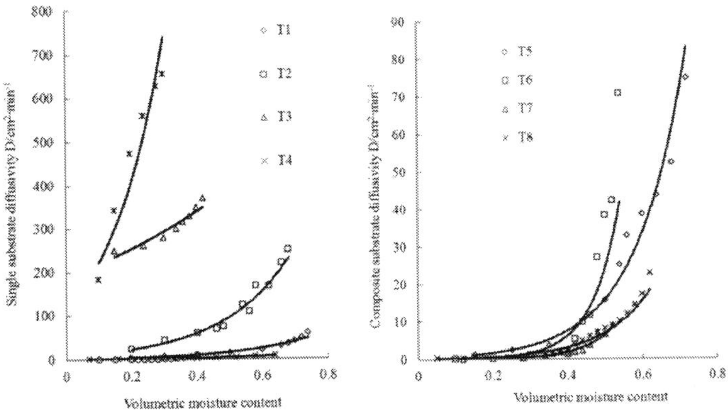

Figure 13. Correlation curve between diffusivity and water content

As Table 15 shown, fit the relation curve of diffusivity D and moisture content θ to get the formula of diffusivity D and moisture content θ.

No.	Fitting relation	Determination coefficient R^2
T1	$D(\theta)=1.6013e^{4.5724\theta}$	0.9771
T2	$D(\theta)=8.5301e^{4.8491\theta}$	0.9732
T3	$D(\theta)=188.64e^{1.4685\theta}$	0.9051
T4	$D(\theta)=0.6172e^{3.8262\theta}$	0.9804
T5	$D(\theta)=0.3662e^{7.5308\theta}$	0.9844

No.	Fitting relation	Determination coefficient R^2
T6	$D(\theta)=0.0256e^{13.694\theta}$	0.9050
T7	$D(\theta)=0.013e^{12.3835\theta}$	0.9823
T8	$D(\theta)=0.1142e^{8.186\theta}$	0.9511
CK1	$D(\theta)=122.04e^{6.0059\theta}$	0.9155
CK2	$D(\theta)=0.0034e^{18.611\theta}$	0.9572

Table 15. Comparison of diffusivity fitted formulas

As shown in Table 15, the relationship between the horizontal diffusivity of substrate and moisture content of substrate are all match with the empirical formula $D(\theta)=ae^{b\theta}$, changed as exponential function. The two have highly significant positive correlation relationship.

6.2.5. *Comprehensive analysis*

Comprehensive assessment is performed to the 8 tested substrates with the matrix method, the standard is the larger total porosity (70%~90%), the higher score; the water-air ratio close to 3.0 is the best; because the substrate is hard to hold water, easy to leak after irrigation, the lower permeability coefficient, the higher score; for water availability, the larger substrate available water content, the higher score; since the diffusivity may affect water absorption by crop, the larger diffusivity, the higher score. Finally, comprehensive assessment is performed according to the synthesis score to various substrates; refer to Table 16.

Influence factor	T1	T2	T3	T4	T5	T6	T7	T8
Total porosity	3	4	7	5	1	6	2	8
Water-air ratio	4	2	1	3	8	5	6	7
permeability coefficient	3	1	5	2	6	8	7	4
Water availability	6	3	1	2	5	7	4	8
diffusivity	6	7	8	4	5	2	1	3
Synthesis score	22	17	22	16	25	28	20	30

Table 16. Matrix table

Accordingly, the synthesis scores of T6 and T8 are higher than other substrates obviously, which means T6 and T8 are better than others. Viz. the total porosity of T6 and T9 are larger, with good air permeability, reasonable water-air ratio, higher available water content, and in favor of crop absorption. Though the permeability coefficient and diffusivity of T8 are smaller, the irrigation evenness can be controlled by adjusting the irrigation frequency and irrigation amount.

6.3. Conclusions

1. By comparing the porosities of 8 substrates, the substrate formed by Domestic peat, vermiculite and perlite in a mass ratio of 3:1:1 has the largest total porosity, 84.7%; its water-air ratio is 2.9 with good air permeability and reasonable water-air ratio.

2. By comparing the permeability coefficients, the permeability coefficient of single substrate is higher than the composite one. When a single substrate is selected for as the culture and breeding substrate, please pay attention to adjust the irrigation frequency reasonably.

3. By comparing the 8 substrates, the available water and easy available water capacity of the substrate with a mass ratio of 3:1:1 among domestic peat, vermiculite and perlite, and the one with a mass ratio of 2:1 between domestic peat and perlite are the highest, viz. the absorbable water for crop is high.

4. The relationship between the horizontal diffusivity of substrate and the moisture content of substrate are all match with the empirical formula $D(\theta)=ae^{b\theta}$, changed as exponential function. The two have highly significant positive correlation relationship.

5. According to the comprehensive assessment with the matrix method, the substrate mixed with domestic peat and perlite in a mass ratio of 2:1, and the substrate mixed with domestic peat, vermiculite and perlite in a mass ratio of 3:1:1 are the best. The total porosity of the two substrates are larger, with good air permeability, reasonable water-air ratio, higher available water content, and in favor of crop absorption, so they may be promoted and applied in production.

Author details

Fusheng Ma[1*] and Haiyan Fan[2]

*Address all correspondence to: mfs@bwsti.com

1 Beijing Hydraulic Research Institute, Beijing, China

2 Beijing Engineering Technique Research Center for Exploration and Utilization of Non-Conventional Water Resources and Water Use Efficiency, Beijing, China

References

[1] Key soft subject research project of Beijing Municipal Science & Technology Commission, "Water-saving strategy research in suburb of Beijing", (Subject No.: Z0005170040231), Technical report [R]. Beijing Institute of Water Conservancy Scientific Research, 2007.

[2] Guo Shenghu, Kong Dejie, Zhang Yuanpei, et al.,"Effects of different irrigation amounts on water consumption and water utilization efficiency of greenhouse cucumber[J]". *Journal of Anhui Agricultural Sciences*, 2010, 36(34): 20568-20570. (In Chinese with English abstract)

[3] Shen Xiaojun, Sun Jingsheng, Zhang Jiyang et al., "Research on water consumption law of winter wheat under deficit irrigation [J]". *Yellow River*, 2007, 29(11): 68-70.

[4] Zheng Guobao, Kong Dejie, Zhang Yuanpei, et al. Effects of drip irrigation under film on water consumption of pepper in solar greenhouse[J]. Journal of Agricultural Sciences, 2010, 31(4): 50-52. (in Chinese with English abstract)

[5] Zhao Ying, Li Bo, Wang Tieliang, et al. Study on water consumption of greenhouse eggplant under small tube flow irrigation[J]. Water Saving Irrigation, 2011(3): 33-35. (in Chinese with English abstract)

[6] Tolga Erdem, Levent Arm, Yesim Erdem, et al. Yield and quality response of drip irrigated broccoli(*Brassica oleracea L. var. Italica*) under different irrigation regimes, nitrogen applications and cultivation periods[J]. Agricultural Water Management, 2010(97): 681-688.

[7] Zhao Jinghua, Hong Ming, Ma Yingjie, et al. Journal of Irrigation and Drainage, 2010, 29(5): 94-97. (in Chinese with English abstract)

[8] Liu Hongguang, He Xinlin, Wang Yaqin. Effects of regulated deficit irrigation on water consumption and yield of grape[J]. Journal of Irrigation and Drainage, 2010, 29(6): 109-111. (in Chinese with English abstract)

[9] Abrisqueta J M, Mounzer O I, Ruiz-Sánchez M C. Root dynamics of peach trees submitted to partial root-zone drying and continuous def-icit irrigation[J]. Agricultural Water Management, 2008, 95(8): 959-967.

[10] Gu Zhonghou, Tian Youliang, Guo Liansheng. Study on the sap flow dynamics and transpiration regulation of pinus tabulaeformis in Daqing mountains[J]. Forest Resources Management, 2006(6): 57-59. (in Chinese with English abstract)

[11] Philips N, Oren R, Zimmermann R. Radial patterns of xylem sap flow in none diffuse and ring porous tree species[J]. Plant Cell Environ, 1996(19): 983-990.

[12] Sun Huizhen, Zhou Xiaofeng, Kang Shaozhong. Research advance in application of heat technique in studying stem sap flow[J]. Chinese Journal of Applied Ecology, 2004, 15(6): 1074-1078. (in Chinese with English abstract)

[13] Ma Fusheng, Liu Honglu, Wu Wenyong, et al.Water consumption regularity of soilless culture Anthurium irrigated by drip irrigation in greenhouse[J].Journal of irrigation and darainge, 2011,30(6):61~64.

[14] Jiang Desheng. *Modern horticultural media* [M]. Beijing: China Forestry Publishing House, 2006: 173-186.

[15] Karam F, Saliba R, Skaf S, et al. Yield and water use of eggplants(*Solanum melongena L.*) under full and deficit irrigation regimes[J]. Agricultural Water Management, 2011(98): 1307-1316.

[16] Guo Yuanyu as the chief editor. *Irrigation and drainage engineering* (Edition 3) [M]. Beijing: China WaterPower Press, 2007.

[17] GB/T 50485-2009, *Technical code for microirrigation engineering*[S].

[18] Qi Haiying, Zhang Shuxia, Song Chaoyu, et al. Effects of different substrates and watering modes on physicochemical properties of substrate and ornamental quality of Euphorbia pulcherrima[J]. Journal of Plant Resources and Environment, 2009, 18(3): 74-80. (in Chinese with English abstract)

[19] Yang Wenbin, Hao Zhongyong, Wang Fengxin, et al. Effect of different irrigation thresholds on growth and yield of garland chrysanthemum (*Chrysanthemum coronarium L.*) in greenhouse[J]. Transactions of the Chinese Society of Agricultural Engineering (Transactions of the CSAE), 2011, 27(1): 94-98. (in Chinese with English abstract)

[20] Tian Yi, Zhang Yulong, Yu Na, et al. Effect of different low irrigation limit on growth,quality and yield of tomato under subsurface drip irrigation in greenhouse[J]. Agricultural Research in the Arid Areas, 2006, 24(5): 88-92. (in Chinese with English abstract)

[21] Sun Huayin, Kang Shaozhong, Hu Xiaotao, et al. Response of greenhouse sweet pepper under alternate partial root-zone irrigation to different irrigation low limits[J]. Transactions of the Chinese Society of Agricultural Engineering (Transactions of the CSAE), 2008, 24(6): 78-84. (in Chinese with English abstract)

[22] Li Xia, Xie Yingge, Xue Xuzhang, et al. Difference of transpiration, fresh matter accumulation and fruit yield for potted tomato under moisture content of medium[J]. Acta Horticulturae Sinica, 2010, 37(5): 805-810. (in Chinese with English abstract)

[23] Steady and health development of flower industry in China – 2009 analysis of statistical data for flower in China. Refer to the website of State Forestry Administration. http://slzy.forestry.gov.cn/portal/main/s/2437/content-438856.html

[24] Poole R T, Conover C A. Fertilizer levels and medium affect foliage plant growth in an ebb and flow irrigation system[J]. J. Environ. Hort, 1992, 10(2): 81-86.

[25] Erin James, Mac van Iersel. Ebb and flow procuction of Petunias and Begonias as affected by fertilizers with different phosphorus content[J]. HortScience, 2001, 36(2): 282-285.

[26] Catherine A. Neal, Richard W. Henley. Water use use and runoff comparisons of greenhouse irrigation systems[J]. Proc. Fla. State Hort. Soc, 1992(105): 191-194.

[27] Chase R, Conover C A. Algae control in an ebb and flow irrigation system[J]. Proc. Fla. Stete Hort. Soc., 1993(106): 280-282.

[28] Robert W. Rigsby. Ebb and flood irrigation system[P]. United States Patent: 5307589, May 3, 1994.

[29] Ren Jianhua. "Water and nutrient solution ebb-and-flow irrigation"[J]. *Water-saving Irrigation*, 2004(3): 49-50.

[30] Yang Tieshun. "Discussion on surface ebb-and-flow irrigation"[J]. *Agriculture Engineering Technology* (Greenhouse Horticulture) , 2009(4): 23-24.

[31] Zhao Chanli, Gao Yuting. "New ebb-and-flow irrigation and cultivation technology settled in Ningxia"[N], *Huaxing Times*. August 19, 2010. Edition 002.

[32] Zhang Xiaowen, Tian Zhen, Liu Wenxi, et al. Research and extension of Ebb-and-flow irrigation system[J]. Transactions of the Chinese Society of Agricultural Engineering (Transactions of the CSAE), 2011, 1(1): 80-83. (in Chinese with English abstract)

[33] Zhang Li, Wang Yong. Primary research of tidal irrigation potted *Hydrangea macrophylla*(Thunb.) seringe[J]. Northern Horticulture, 2011(20): 77-79. (in Chinesewith English abstract)

[34] GB/T 50485-2009, *Technical code for microirrigation engineering*[S].

[35] GB/T 50485-2009, Technical code for microirrigation engineering[S].

[36] Li Mingsi, Kang Shaozhong, Sun Haiyan. Relationships between dripper discharge and soil wetting pattern for dripirrigation[J]. Transactions of the Chinese Society of Agricultural Engineering (Transactions of the CSAE), 2006, 22(4): 32-35. (in Chinese with English abstract)

[37] John M. Dole, Janet C Cole, Sharon L von Broembsen. Growth of Poinsettias, nutrient leaching, and water-use efficiency respond toirrigation methods[J]. Hortscience, 1994, 29(8): 858-864.

[38] Daniel I. Leskovar. Root and shoot modification by irrigation[J]. HortTechnology, 1998, 8(4): 510-514.

[39] Meng Qingguo, Chen Fengming, Liu Yingjie. "Application of drip arrow microirrigation technology on pottedflower production" [J]. *China Flower & Horticulture*, 2005, (14): 50-52.

[40] Pang Yun, Zhang Junsheng, Zhang Zhi, Zhang Wenzhu. "Comparative test between various water-fertilizer treatment and soil culture border irrigation for soilless culture tomato under drip irrigation"[J]. *Acta Agriculturae Boreali-sinica*, 2006,21 (Special) : 41-43.

[41] Yue Bin, Wang Liuyun. "Test research on drip irrigation soilless culture tomato"[J]. *Water-saving Irrigation*, 1998(2): 18-1948.

[42] Wang Ronglian, Yu Jian, Tan Yumei et al. "Influence to yield of soilless culture cherry and tomato from water and fertilizer coupling during drip irrigation in greenhouse"[J]. *Journal of Irrigation and Drainage*, 2009, 28(4): 87-89.

[43] M.K. Merica, I.H. Tuzelb, Y. Tuzelc, G.B. Oztekinc. Effects of nutrition systems and irrigation programs on tomato in soilless culture [J]. Agricultural Water Management, 2011(99): 19-25.

[44] D. Massaa, L. Incroccib, R. Magginib, et al.Strategies to decrease water drainage and nitrate emission fromsoilless cultures of greenhouse tomato[J]. Agricultural Water Management, 2010(97): 971–980.

[45] Zhong Gangqiong, Chen Yongbo, Long Jun et al., "Research on irrigation amount and irrigation mode for konjac soilless culture"[J]. *Agricultural Engineering Science,* 2005,21(3):340,341&353.

[46] Wang Xiukang, Li Yuannong, Xing Yingying et al. "Influence from dripper flow to summer maize root system"[J].*China Rural Water and Hydropower,* 2010 (11) :66-68.

[47] Tian Jilin, Wang Yinhu. Current situations and prospects of researches on soilless-culture substrates[J]. Acta Agriculturae Shanghai,2000,16 (4):87~92

[48] Li Shijun, Gao Lihong, Zhuang Zhonglian. Achtievement new technology of the recent past in soilless culture in China and its trend[J].Journal of Changjiang Vegetables,1997,(5):1~4.

[49] De Boodt M, Verdonck O. The Physical Properties of the Substrates in Hort[J] . Acta Hort, 1983, 26: 37~44.

[50] Zhou Yanli, Cheng Zhihui, Meng Huanwen, et al. Effects of organic substrate compositions on growth yield and quality of tomato[J].Jour. of Northwest Sci-Tech Univ. of Agri. And For.:Nat.Sci.Ed., 2005,33(1):79~82

[51] Li Jingrui, Zhang Tiejun, Gao Hongbo,et al.Effects of different substrates on growth and chlorophyll fluorescence characters of eggplant seedlings[J].Acta Agriculturae Boreali-occidental is Sinica, 2010,19(6):139~143.

[52] Wang Yajun, Wei Xinghu, Xie Zhongkui, et al.Effects of different culture medium on growth and bulb of cutting flower lily[J]. Acta Agriculturae Boreali-occidental is Sinica, 2003,12(4):109~112.

[53] Zhang Jianhua, Deng Pu. Effects of different culture medium on survival rate and growing situation of anthurium and raeanum tissue culture seedling after transplanting[J]. Journal of Shanghai Jiaotong University:Agricultural Science, 2009,28(6): 624~626(in china)

[54] Zhang Ning, Feng Mei. Ping Jicheng. Effect of different substrates on growth of strawberry[J].Northern Horticulture, 2011,(16):51~53. (in chinese)

[55] R. Kasten Dumroese, Juha Heiskanen, Karl Englund,et al. Pelleted biochar: Chemical and physical properties show potential use as a substrate in container nurseries[J]. Biomass and Bioenergy, 2011,(35):2018~2027.

[56] David W R. Water, media and nutrition for greenhouse crops[M]. Batavia,Illinois USA: Ball Publishing, 1996.

[57] Tang Jing, Kang Hongmei.Physical and chemical properties of several important media[J].Chinese Journal of Soil Science,2006,37(2):291-293(in chinese)

[58] Zhao Jiuzhou, Chen Jiemin, Wang Yiqing, et al.Studies on the physical and chemical characters of six-soilless-media[J].Journal of Laiyang Agricultural College, 2001,18(3): 161~164,173. (in chinese)

[59] Li Jing, Zhao Xiulan, Wei Shiqiang, et al.Study on the physico-chemical properties of soilless cultural substrates of pollution-free vegetable[J].Journal of Southwest Agricultural University, 2000, 22(2):112~115. (in chinese)

[60] Chen Zhende, He Jinming, Huang Junjie, et al. Composition and Physical and Chemical Characteristics of Plug Seedling Media of Vegetables[J].Transaction of the CSAE, 1998,14(2):192~197. (in chinese)

[61] Tian Jilin, Xi Zhenbang, Chen Chunhong. Study on the feasible parameter of substrate quality for soilless culture[J].Acta Agriculturae Shanghai, 2003,19(1):46~49. (in chinese)

[62] Chen Lixin. *Coursebook for Soil Experiment and Practice*[M].Harbin: Northeast Forestry University Press, 2005: 59~63.

[63] Duan Zhaolin. Application on the matrix of environmental impact comprehensive assessment based on the principle of getting value from environmental elements proportion value[J].Yunnan Environment Science, 2003,22(4):59~62. (in chinese)

[64] Li Douzheng. *Study on effects of components and particle sizes on pore properties of media*[D]. Shandong Agricultural Unverisity, 2006. (in chinese)

[65] Du Yanling, Xu Guo'an. *Finite element method and electric network method for seepage analysis*[M].Beijing: Water Conservancy and Electric Power Press, 1992.

[66] Shi Lianhui, Zhang Zhiguo Liu Dengguo et al. Comparison of physiochemical properties between spent mushroom compost and peat substrate and adjustment[J]. Transaction of the CSAE,2008,24(4):199~203. (in chinese)

[67] Yao Zhenxian, Zhang Dapeng, Yao Miaosen. Breslers's mathematical model for determination distance between emitters by soil hydraulic parameter[J]. Transaction of the CSAE, 2012, 2(2):65~68. (in chinese)

[68] Jiang Shufang, Wan Shuqin, Kang Yuehu.Effects of soil matric potential on Chinese cabbage yield and water use efficiency under drip irrigation[J].Water Saving Irrigation, 2009,(10):20~22. (in chinese)

[69] Zhang Chao, Kang Yuehu. Wang Shuqin, et al.Effects of soil matric potential on cowpea yield and irrigation water use efficiency under drip irrigation[J].Journal of Irrigation and Drainage,2010,29(4):30~33. (in chinese)

[70] Wang Shuqin, Kang Yuehu, Liu Shiping.Effects of soil matric potential and topping on tomato growth and water use efficiency under drip irrigation[J]. Journal of Irrigation and Drainage, 2009,28(2):1~4. (in chinese)

[71] Jiang Shufang, Wang Shuqin, Kang Yuehu, et al. Effects of soil potential on lettuce growth and water use efficiency under drip irrigation in semi-humid regions of north China[J].Journal of Water Resources and Water Engineering, 2011,22(6):72~76.

[72] Theodore W Sammis. Comparison of sprinkler, trickle, subsurface, and furrow irrigation methods forrow crops[J]. Agron J, 1980,72:701~704.

[73] Jia Junzhu, Kang Yuehu, Wang Shuqin, et al. Effect of soil matric potential on lycium barbarbum L. growth under drip-irrigation[J]. Journal of Irrigation and Drainage, 2011,30(6):81~84.

[74] Rivière L M, Caron J. Research on substrates: state of the art and need for the coming 10 years[J]. Acta Hort, 2001,548:29~42.

[75] Fan Yanwei.Study on numerical simulation of interference infiltration characteristic in film hole irrigation[D]. Yangling: Northwest Agriculture and ForestUniversity, 2008.